Why Don't We Drive *from* Portland, Oregon, *to* Argentina?

Constance Glidden Josef

ISBN 978-1-63961-122-5 (paperback)
ISBN 978-1-63961-123-2 (digital)

Copyright © 2021 by Constance Glidden Josef

All rights reserved. No part of this publication may be reproduced, distributed, or transmitted in any form or by any means, including photocopying, recording, or other electronic or mechanical methods without the prior written permission of the publisher. For permission requests, solicit the publisher via the address below.

Christian Faith Publishing, Inc.
832 Park Avenue
Meadville, PA 16335
www.christianfaithpublishing.com

Printed in the United States of America

Dedication

To the love of my life, dearest friend and partner forever. Thankfully a persistent man who encouraged me for 20 years to pull out all of the material from an old warn out bag and publish this enduring adventure story. To Conrad…my love forever. Thank you.

Thank you to all of the wonderful people throughout the Americas who welcomed us and often took us into their homes and made us feel like family. There were always smiles and wonder in every city and town we travelled through.

Finally, Velveeta. The journey could never have taken place if it had been for her marvelous workings. Purchased at the age of 143,000 miles everyone said she wouldn't make it from Portland, Oregon to Argentina. I knew she could and she would. Lots of tender loving care and attention before and during the journey. She never faltered. Velveeta was our home on wheels for over 12,000 miles. A remarkable 4-Runner who now lives in Argentina with our favorite adopted family.

Introduction

If I told you a story about a forty-four-year-old woman driving from Portland, Oregon, to South America in a 4x4 Toyota, you would think of another adventure/travel story. Nope, it's a romantic, zany adventure, and, "Well, why not?" You couldn't in your wildest dreams make this up. It's better than girls going shopping, lunching, golfing, or out for a skiing weekend. After years in the same profession in damn good jobs (one example being United States' Embassy in Egypt for two and a half years), she was in the right place at the right time with good opportunities and responsibilities. A girl from Indiana with her own student loans and fellowships made it thru university and graduate school. With echoes from her mother saying, "Get a husband, girls don't need college."

If, like her, your parents did not have money, nor give you a silver spoon. You were on your own. All the ERA (Equal Rights) commitments could not get you a free pass to anything. She persevered and found a profession she loved—information technology (IT). Yep, a geek but not a coder, but rather an analyst, planner, and team member. She got her first taste of IT while as a graduate assistant with a fellowship drawing her attention into the world of computer software.

The story is going to share what happened to her and her good friend. It's real, funny, romantic, and in this story, you might even find yourself crying, laughing, winching, and possibly losing your breath. The story shares adventures of border crossings, guards demanding money, and toilets with no seats.

Here is a hint of what is to come: a) motels with the exact same landscape pictures on the hotel's entry walls; b) showers with no showerheads and signs reading: water turned on every other third day and hot water only on Sundays; c) border crossing with guards having unending pockets for money to pass; d) checkpoint guards on roads who have lots and barns full of impounded cars and trucks always demanding more money, and not in their country's currency but American greenbacks.

Now, don't think this is all bad. There are far more stories of great fun, exploration, breathtaking sights, and endless warm, friendly, and welcoming people in every country.

Why you ask would she even contemplate such an adventure? Let's just say she was ready. Although driving from Portland, Oregon, to Argentina never really occurred to her until a suggestion came from a friend over dinner and drinks.

They met and immediately hit it off. He was a Brit who lived in the good ole US of A for many years, flying helicopters on the west coast after leaving South America. He continuously felt the pull of his South American flying adventures. He lost his helicopter company in South America, and he continued to feel the loss even after twelve years. So, after many stories, she popped the question back to him. Why not drive to South America? Drive from Portland, Oregon, to Argentina.

I write this introduction, having known and lived with her for over twenty years. She is my best friend, a partner in life, and wife. Getting her to write and publish these stories was not easy. The adventures took place exactly twenty-five years ago this year of 2020, back in 1995. It's time.

Now that you have a clue as to how the idea to drive from Portland, Oregon, to Argentina came about, I (Constance) want to give you some additional background.

Although I traveled, for what I thought was extensive, this adventure would always remain on top as providing surprises almost every day. While traveling through the Americas, I found myself thinking it would be nice to have a relatively quiet day on the road, then BAM! The quiet was always an introduction to the surprises that would appear just around the bend in the road or in the small towns where we would spend our night.

Let me provide a brief background of myself as a starting point.

As mentioned, I was forty-four at the time we set off from Portland in April 1995. This adventure was not a midlife crisis event. Although I was yearning for something really new for an adventure. Now, that is not to say I was only fixated on my career. I'm from a family of modest means, but my father would have us in the station wagon at least once a month for an outing somewhere. So I learned to enjoy exploring and discovering new places.

I found the same excitement in attending college and graduate school. While in graduate school, my fellowship introduced me to my dream career and also helped me to pick up a few side hobbies such as flying and photography. My photography leads me to discover the great beauty of the Pacific Northwest through the many photography magazines and flying itineraries. Although I was offered paid tuition for law school, I turned it down and started working into making a dream become a reality in an information technology (IT) career.

I started my planning for my first job out of graduate school by subscribing to Portland's Sunday paper and reading the want ads. Reading about the different companies and position openings opened my eyes to moving to the Pacific Northwest and finally have an exciting career at the age of thirty-three. This led me to great opportunities with Nike and EDS. Those opportunities eventually lead me to go out and win consulting and teaching positions in Cairo, Egypt, for two and a half years. There, I pursued adventures across Egypt, Europe, England, and Greek Islands. It was as if life in Cairo awoke my soul and spirit. A combination of being in the right place at the right time and destiny.

I returned to Portland, and consulting opportunities were present and took me all across the country. I loved my work and focused on details of systems and operations.

My girlfriends from across many years were still with me; and we never stopped talking, writing, and having fun. I was out skiing, flying, and photography on weekends in Colorado, paragliding in Southern Oregon, golfing and weekend beachside retreats with girlfriends, plus enjoying the Portland art's talent (which there were many choices classical, pop, and jazz). I even took an opportunity to drive to the east coast, live for three to four months, and then drive back to the great Pacific Northwest.

Guys were not to be the focal point of my life. I already had two unsuccessful marriages and didn't need to find myself in those circumstances again. Not even close.

So what happened to cause me to consider this South American adventure? Gosh, I met a fellow who talked of new possibilities that I had not thought about. His adventures, at times, were similar but oh so different from mine. He knew Africa (but locations I had not visited), Asia, and all of South America. He was a Brit who spoke excellent Spanish because his business, flying planes and helicopters, led him to many different opportunities and businesses in South America where he lived for twelve years. He knew South America from the air and loved it. He told me so many stories. So I put the question out there.

Why not drive from Portland, Oregon, to Argentina?

So it was that we sat down and started discussing just what would have to occur in order to put this adventure together to make it work. We then engaged in a yearlong ambition of planning, organizing, and moving our lives in a totally different direction. Everything would be changed. Even the aspect of planning brought me into whirls of excitement, but never losing focus on my work, which would eventually take a back seat (so to speak) to this adventure of the Americas.

Chapter 1

From Portland Onward

From Portland, my daily ritual was to write the happenings of the day, every day. That philosophy would change over time, but it seemed like a good idea at the outset because I thought we might forget the *specialness* of each new location. Each day did become mixed with elements of awkwardness and interruptions that were *off the wall*. People, places, and things we often might take for granted in what may be termed *our typical* daily life. We came to learn to *take nothing for granted and always, always plan your route ahead of time and always stay alert.*

Closing down *home,* packing up, and leaving all behind is more difficult than I thought. I have moved several times, but this was so different. If you didn't store it away for a later day and you don't need it on the trip, pitch it. We were going up and downstairs, taking bits and pieces to load Velveeta (our Toyota Four Runner) in the final moments.

I was throwing things down the trash shut like mad. Especially, when I found out the new refuse disposal had an automatic compressor. So when I could not find a way to get rid of my .22 long rifle through the police, I tossed it down and mangled it up! Why on earth we didn't have an explosion after dumping all of the bullets into the bin is a surprise to me. But there we are, and the job was accomplished! Velveeta was so badly loaded. She leaned to one side, but it was late and time to leave. A friend gave us her home for an overnight stay before leaving Portland for the Oregon coast and our road South. We fell into bed, exhausted.

The next day, we drove over to the coastal town, Lincoln City. It was pouring, and I mean pouring rain. Velveeta was too heavy in the back and on one side. We were able to get to the beach house. Still pouring, but fortunately, Velveeta could spend the night in the garage. We unpacked everything, then went through and downsized, leaving behind items we felt would not really be needed, such as a picnic basket full of dinnerware and our gazebo. Really, we packed like we were going to the beach for a weekend picnic rather than to Argentina through the Americas. Dear Velveeta was heavy and *badly* loaded. We knew we would need to unpack and reposition everything as we traveled.

Finally, we were able to depart Oregon on the morning of April 28, 1995. A day or so, late in departing, but we needed more time to organize and check our map bearings.

The drive down Oregon's Coast from Lincoln City

The drive down through California had scenery that was fabulous. I drove the whole time. (Hogged the wheel, so to speak. I love to drive.) On day two, we covered close to three hundred miles. On day three, we repacked Velveeta and distributed the load a little better, although still not balanced.

On Monday, May 1, we pulled into Patterson, California. This was a gift to Robbie. It was his lovely treat to see his friends Ray, Bonnie, and Sue, after thirty years. They had worked together in Robbie's helicopter business. Everyone got a bit teary-eyed, so we only stayed a couple of hours. It would have been nice to stay overnight, but Robbie thought it better to carry on. It was wonderful to see them and listen to the many stories of those thirty years ago. And for me, it was also a treat to hear the stories I wondered were just Robbie's stories. Now, I learned they were real-life experiences.

Wind turbines through Northern California

That night, we stayed in Fresno. Driving through the hills of California brought the sight of so many wind machines. There seemed to be thousands on the hillsides. It was incredible, and once again, I have not taken lots of photographs. The roads do not allow for just the right shot. I captured what I could, hanging out the window, and it was a spectacular sight. Throughout the San Joaquin Valley, we saw dairy cattle, rows and rows of vineyards, and many wineries call out our name to take a pause and sip wine and take along a few bottles for the road. But we resisted the temptation and kept to our plan.

We arrived in Las Vegas the next day (May 2). And at long last, we had sunshine. The night in Vegas was actually a quiet one. Ninety degrees and very dry. We didn't really do Vegas justice. The difference between *the strip* and *old town* was in itself entertaining. We couldn't get in the mood to enjoy because we were on our way to Mexico. It was time to check the maps, look over Velveeta, and say goodnight.

The next day, we were over the Hoover Dam. The road engineering was absolutely spectacular. If you haven't seen this firsthand, you really must. I could not believe what it must have taken to design and build. Such a height and such a span of water that is being held back and controlled through the dam. A great number of people and their dangerous jobs and lives were lost to deliver an incredible structure that looked one with nature, although you knew it wasn't.

The drive became boring after leaving the dam. We spent the night in Williams, Arizona. The next day, we were off to the Grand Canyon by train ride. It was fun with the cowboys and their play revolvers providing a train hold-up.

Upon arrival at the canyon, I quickly walked to the canyon's edge. This was my first experience standing at the edge of the Grand Canyon. And I thought the Hoover Dam was overwhelming. Hoover was just overtaken by the Grand! I had flown over this magnificent carving into the earth a few times. Now, I was hanging over the canyon's edge with my camera and just stayed in a state of awe over its beauty. It must be splendid to watch its color vary throughout the day, every day, and every season.

The time came to drive on. From Williams to Flagstaff, the weather became very unusual. We had only driven sixty-six miles, but the weather was miserable. Once in Flagstaff, we had two days of snow. Oh, well, take advantage of the *'snow days'* and wash clothes.

The morning of May 7 arrived, along with a temperature of seventeen degrees and another snowstorm. But it was time to go. Driving through the buttes was lovely. Finally, out of the snow and rain, and on to Texas. But Texas also had extremely bad weather. We agreed to change our route and continued our route through Arizona and into Tucson. So, 268 miles that day, but we needed good weather. So a night was spent in Tucson to plan for our entry into Mexico.

Knowing our next stop would be Mexico, we took time to give Velveeta attention. An oil change and general maintenance check before rolling into unknown available care for Velveeta were called for.

We prepared our paperwork for the border crossing with Sanborns. We didn't want problems in crossing the border at Nogales. Sanborn prepared papers, gave us advice, and helped to relieve some of my worries. When we reached the Mexican border, no problems. There was a very surly Mexican official who looked at the two passports—one with the name Hilliker (my name) and Robinson (Robbie's) on the other. The officer looked across the two of us, stamped each passport, and on we went.

Amazing what a stamp can do for your health. I could breathe again. That accomplishment only took just over 2,200 miles to achieve! I wonder what the rest of the ten thousand plus miles of wonder will bring.

Travels in United States							
Date	Mileage		Number Miles	Location for the Night (or more)	Notes	Days Travelling in Country	Miles Travelled
4/25/1995	144772	144851	79	Portland, Oregon to Lincoln City OR	Left Portland for Argentina	USA 15 Days	2011
26-Apr	144851	144910	59	Lincoln City, OR	Stayed a couple of days at Marty's home in order to balance everything packed in Velveeta. Pouring rain for both days.		
28-Apr	145119		209	Gold Beach, OR			
29-Apr	145,412		293	Ukiah, CA			
30-Apr	145545		133	Vallejo, CA			
1-May	145731		186	Fresno, CA			
2-May	146149		418	Las Vegas, NV			
5/3/- 5/4, 1995	146372		223	Williams, NV			
5/5/- 5/6, 1995	146438		66	Flagstaff, AZ	It's SNOWING!		
5/7/- 5/9, 1995	146706		268	Tucson, AZ	Business names: Cluck U Chicken, Dirt Bag, Greasy Tony's - all restaurants!		
10-May	146783		77	Nogalos, MX Borader	USA into Mexico Border Crossing		

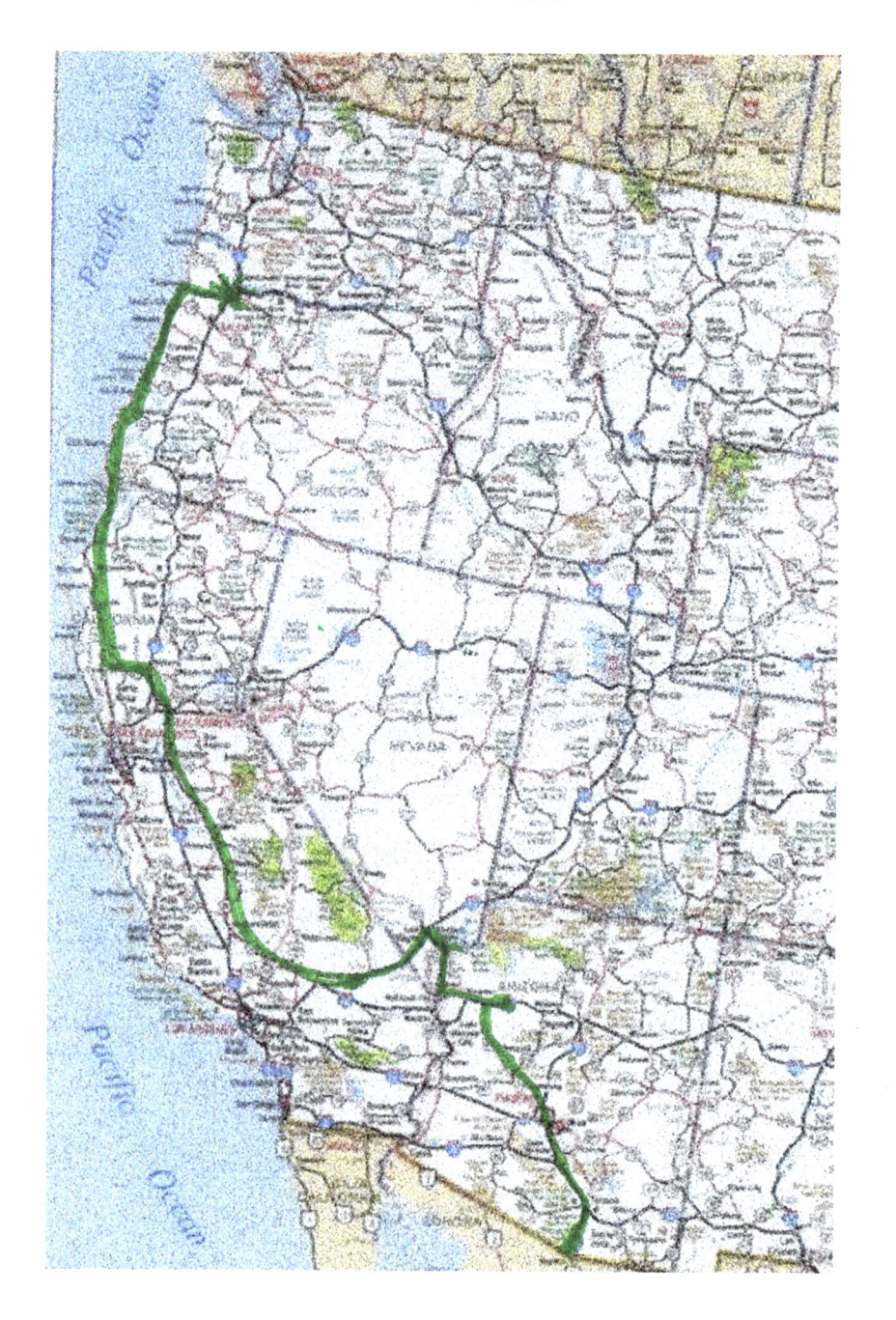

Route Map thru Western United States

Chapter 2

Mexico, the Land of Beauty

The passing of the days of the week disappeared, and I began to no longer think about what the calendar displayed. Rather, our road maps, reference books, guidebooks of town center maps, Inns, and ATMs would be our calendar.

After leaving the Mexican border, we stayed our first night in Mexico in the city of Hermosillo. The next day, May 11, it was on to a little town in the interior called Yecora. Not a far distance, just 170 miles.

Sitting in a cool room in the only Inn in Yecora, my thoughts and reflections of our drive from Portland, Oregon, to this point were filled with excitement. I thought mostly of just having packed up and left. The sight of everything pulled out of Velveeta in order to have what we needed at hand and that which wasn't ever needed—gone. Velveeta was balanced in weight and mechanically ready for the vast drive. Who would have imagined that Velveeta, at 143,000 miles, was too old for this adventure? We certainly didn't! No matter what the age, TLC was the youthful answer. That goes for us as well!

Sierra Madre

The drive, on Highway 16, took us over the Sierra Madre. At times, I found myself amazed at the sights. Beautiful rolling hills, brown from the wait for rain. The rain seemed to hold itself and collected within a gigantic canyon. Looking out, it was as if you looked upon a rim going around many lovely *volcanic*-shaped mountains. The color of copper is prominent. And at one point, the rock looks as though it had taken on the patina color, green that comes from the weathering of copper. Beautiful colors. The road was in very good condition, with inclines and turns throughout the route. There are very few spots to pull off and take photos, but I managed a couple. I drove Velveeta the entire route. She still continues to perform very well. At times, I wanted to take photos of the beautiful view, but I can't do it justice. You must experience the drive to really capture the moments of sheer beauty.

I find that I'm still not at a point of relaxation, yet I have not stepped foot into an office nor thought about work since March 31. I also don't intend on doing so for some time!

Robbie in all smiles

Robbie is so full of life these days. Not that he wasn't before, but it is different. He smiles from within—laughs come from his toes. He was excited about returning to his beloved adopted South American home.

Getting back to the village of Yecora, and especially the inn. It was clean and very inexpensive, but it isn't a Motel 6 and you can't really use the shower. The water doesn't drain, reminding me of a hotel in the middle of the Egyptian Western Desert. It was built for the visit of a Canadian ambassador. It all worked fine while he was there for the first visit, but after that, well, don't try it. Although, I do admit it made for an added adventure on the way to Siwa to locate the Temple of the Oracle of Amun. But off subject, since Siwa was an earlier stunning adventure in my life a few years earlier, but a good comparison!

Our evening is spent working out a route for tomorrow. I hope the mountains keep us cool. I'm most pleased that I'm no longer with a sick stomach. Finally!

Mexico Cactus in Bloom

The scenery continues to be absolutely magnificent. We crossed the scrub-filled valley and started ascending into the Sierra Madre's (ten thousand feet) and its cactus that are all in full bloom. One cannot describe the magnificence, the grandeur of driving up through the mountains. I wish I had a sporty Porsche to drive, but I had lots of fun with Velveeta. We dropped from Yecora to Cuantemoc. It is a very rich country. Not a single pull-off, whatsoever, to take photographs of this scenery. It is a great pity. I just slow down and swing the camera out the window. I know I could stop driving and give in to Robbie, but no, I'm loving this drive.

The vistas are limitless in the distance, as well as trees, cactus, and copper canyons. Downshifting through Velveeta's gears along with her superb performance really gave me a chance to pretend to be driving in a sporty manner. The appearance of slow-moving trucks was not a drawback of the excitement but rather an increase of good sporting fun. The courteous drivers waved us by when our vision was limited. Those who thought to ignore received a quick press of the mighty hornblower—Robbie's absolute thrill! (He truly believes this was the best investment at a value of $3.29). Hornblower is a can of compressed air, and when you press the top button, the air blows out through a horn. The sound is beyond loud. The sound echoes off of the mountain walls leaving Robbie rolling with laughter in his seat.

Soon, the road begins to straighten and drop down into the valley. The thrill of driving the hairpin curves and racing past 16 wheelers is going to be left behind with the breathtaking vistas. The consolation of driving through the valley brought an immediate relief from the 100-plus-degree temperature. But the two of us look like two disappointed teenagers walking away from an amusement park that was suddenly closed down. Gosh, where is the fun in driving through a valley of straight roads?

Passing through meadows and forests, the road becomes very narrow. The closeness is so opposite of our previous few days of wide-open space. I am beginning to feel slightly claustrophobic from the total coverage of foliage. Thankfully, the feeling starts to fade along with the ping of disappointment.

Our usual conversation of the day is displaced with a wonder lust of *look and see*. The road is quieted by the shadows of the trees. We find ourselves whispering as though we are in a theater just walking in on a performance. Don't want to intrude. Robbie fines a stop to pull over, and we just sit looking out into a meadow to my right. We roll down the windows to feel the slight, cool breeze.

Closely watching, I realize there is a movement of a horse. Almost invisible in the shadows but moving in a full flowing canter. The horse steps into the full sun lite meadow, and there the horse seems to glow from within his deep chestnut color, as does his very handsome rider. He is the first Mexican cowboy I have ever seen. He rides with an expression of joy, pride, and complete control. From within their movement, I can sense the horse would never disturb the gentle flow of this cowboy. Their movement is one of mutual respect and care—they are partners and friends.

The cowboy sits tall, and upon his head is the traditional Mexican cowboy hat. Chestnut color with a brim that is large, turned up in the front, and its cap is more pointed than a typical American cowboy hat. The hat, cowboy's clothes, and the horse all have the chestnut glow. Only the cowboy's black full mustache and the horse's black mane pose a contrast. They are magnificent.

I want to stop them and capture this cowboy and his flowing magnificence. It would make a great photo, but I know this moment will only ever be captured in my memory. I'm selfish—I don't want to stop watching this site, and I definitely do not want to cause them to stop. They have a natural flow that lingers in the way in which they move.

Carefully and quietly watching, I want to anticipate their change in movement. I can see the cowboy looking out to his front with a slight smile and a daydreamed look. It is his private moment. Then, within that same moment, the cowboy slowly turns his head toward me, and his eyes meet mine. Looks pass between us, and introductions are made on the hand of the breeze. I softly smile with an accompanying look of joy, with eyes slowly closing and speaking to me, knowing I have witnessed a private moment. I'm extending my apologies for intruding. His gracious and full smile is instantaneous.

As he continues to ride, his smile is caught within the reflex of his muscles, and his horse turns his head toward me and salutes with a throw of his mane. The welcome is complete. Cowboy and horse never miss a beat of their rhythmic canter. The cowboy's smile and the horse's throw continue as we pass, and I smile in appreciation. No need to wave or call attention of any kind. It is within our meeting eyes that convey the pleasure, warmth, and gratitude of the welcome.

Robbie and I now remember we should be driving and leave the cool shadows of the meadow. Laughter is still filling Velveeta. We are children giddy from unwrapping a surprise gift. We also soberly remember our travels continue beyond our cowboy's land—what a sight, what a thrill. We look at each other with that special inner smile and glint of the eye—attempting to calm our renewed youthful spirit into quiet adult pleasure. It is unanimous with laughter.

Now, we are into Sunday, May 14, and 670 miles from the Mexican border. We spent the night in Chihuahua and discovered the lovely ice cream vendor across from the hotel. We are still stunned by the cost of the rooms and food in Chihuahua. The ice cream, well, that was another thing. I cannot get over the number of choices of fruit-flavored ice cream. No chocolate, thank goodness, since I hate chocolate. Every one of those creamy ice-cream pops had its last bit with the fruit piece at the end. Yummy and creamy fresh fruit flavor. Lovely.

The drive down to Jimenez was flat and almost beginning to be boring, then a surprise rolled up behind us. We were being pulled over by a Mexican police officer. Of course, I was driving. The uniformed officer walked up to my side of Velveeta and asked me to roll down the window and leans against Velveeta, such as to show the gun holstered under his jacket. Robbie leaned over and asked what we can do to help. The officer wanted us to unload Velveeta. Robbie got out, joined the officer in the back of Velveeta, and opened the back. Robbie pointed out plastic containers with clothes, various belongings, and the portable potty sitting just at the end. He talked and talked with the officer for nearly thirty minutes. Finally, Robbie closed the back and climbed back inside Velveeta, only to say, "Just drive, don't talk, and don't look back. Drive the speed limit." That was it.

That night, Robbie explained that the officer was looking for drugs and didn't believe Robbie when he explained about our plans and route to Argentina. Finally, looking at the inside of Velveeta, the officer gave in and said, "Be on your way." That would not be the last time we would be stopped.

The next day, May 15, we are off and on to Durango and just over three hundred miles. The temperature is more than hot. It must be well over 100, and there is a constant blow of the wind. Opening the window was like opening an oven door. This will make for a very long day.

The route presented flat plains loaded with cactus in full bloom. Periodically, the plains would open up to the gazing of excellent-looking cattle.

The entry into Durango gave us the immediate impression that we would be spending the night in a lovely town. We found a dear hotel while driving down Main Street called Casablanca. Dear ole Boggy was not there, and they didn't have a pianist anywhere to be seen. Ah, but it was great and a reasonable price. Dear Velveeta was tucked in the back and locked up. The weather all the way down to Durango was hazy from the burning of the crops. The views were spectacular, but we couldn't see well enough to take clear photos. Ah, I didn't even see a Mexican sombrero. We were lucky and so thankful to have the Mexican cowboy introduction a few days earlier.

On May 16, we drive down Highway 40 from Durango to Rosario and the highway becomes Highway 15. We start feeling that we are headed for the Pacific Coast. Highway 40 gave us spectacular road engineering. Views are just wonderful. More spectacular on this side of the Sierra Madre. Unfortunately, there were no photographs due to the haze from the burning.

Some nice points to add regarding today's travel: We came to a village called Buenos Aires, then we came to what is known as the devil's spine. In which two parts of the Sierra Madre are bridged together. At this spot, there was a place to pull off, but the air was so smoke-laden that you could not see anything. It was a great, great pity. After that, we came to the village of Los Angles. So, we beat the speed limit of the Concord. From Buenos Aires to Los Angles in about two hours. Not too bad!

Trucker road race

Another amusing note regarding the many large trucks is we came upon three great big trucks in a convoy. Full of hogs. Four tears of pigs on each truck. Absolutely incredible. There was a Texan in a Chevrolet pickup in front of us (Texas license plate). His wife (we assumed) was seated as close to the passenger door as she could get. We are not sure if she just wanted to jump or hang out the window to possibly throw up. This Texan never would pass those trucks. In the background, her voice is sounding similar to the Wicked Witch of the East and screaming, "No, don't pass. No, don't pass."

As far as we know, he is still in the mountains somewhere. We didn't mind taking our turn and passing those trucks. Besides, they wave you on to let you know there are no vehicles coming down. Even at the corners, they can see better than we can. Besides, this all gave Robbie more of an opportunity to use his trusty hornblower and laugh like crazy while saluting and thanking each driver as we passed.

We settle in a nice little hotel in Rosario for the night. We know the Pacific Ocean is out there close by. Hopefully, tomorrow we will see it. Oh, Velveeta has just had a nice bath and is sparkling clean.

This is life. In the distance, waves can be heard continuously coming in against a beach. Flowing and brushing of tree leaves, a breeze coming off the ocean that has the many palm trees swaying overhead. We drove all of 75 miles from Rosario to Novillero, which is right on the Pacific. We stop, and I'm in need of rest, for I have developed very sore driving arms. Robbie is going to be a gentleman of leisure and drink a gin and tonic and smoke a nice cheroot under the thatch roof of a little cabana. From the high Sierra Madres to the Pacific Ocean beach—a grand drive through an absolutely beautiful country.

We decided to stay one day more and put our laundry in for care. My God, I didn't realize how much laundry we had. We sat down and tried to figure out the last time we had washed our clothes. Goodness, it turned out to be Tucson, Arizona. That seems an eon away. A good repack for Velveeta is necessary and give her another wash. Try to remove as much of the terrible road tar and chewing gum. Can't figure out where the chewing gum came from.

Mexico Piggies Playing

Took some walks on the beach. I even discovered little pigs running along the beach. I have no idea where they call home, but they were so friendly, just like little piggy puppies. They sniff at my feet then go off running down the beach in a piggy sequel. I run with them for a while, then turn back. The little piggies really don't want to stop playing.

What I haven't mentioned is the prices in this little village. The room is delightful. The breeze comes in off of the Pacific. It is fifty pesos a night, just less than $10. We had the good fortune of meeting two people from the US of A, Robert and Rosie. They are on their way to Manzanillo. While we were talking about prices in Mexico. Rosie mentions Mexico has never been cheap. They have been traveling in and about Mexico for years. Never cheap. So these guidebooks are for the birds.

We drove from Novillero to Puerto Vallarta, just over 240 miles. Robbie had a chance at the wheel from eight o'clock to five o'clock today. All-day. He rubbed it in, found it all rather exciting! It has been an exceptional day. We came across the most marvelous little hotel. Their sign attracted our attention. I don't know what official name would be given to this type of *inn*. It is full of small villas. Completely, self-contained. Kitchen, three bedrooms, air conditioning. No television. No telephone. It is wonderful and the ocean is within, as one would say, spitting distance. So we decided to stay two days. Marvelous!

We now sit out on the verandah overlooking the ocean in our tiny villa in Puerto Vallarta. Palm trees and Purple Heart flowers. We can see the swimming pool. This is seventh heaven. This is what we dreamed about and the first place we have found. The price is just slightly over $ 30.

Sitting here on a pebble-filled beach, the beautiful Pacific Ocean is constantly lapping within three inches of my stretched-out feet. So tranquil at 6:30 in the evening. The sun is setting; the mountains are standing out clearly on either side. The evening ends with our looking over the map and recognizing tomorrow's route is Manzanillo and it is about a three hundred-mile drive. It should be pleasant down the coast, then a turn into the mountains and back down to the coast.

Pacific Coast

Sunday, May 21, we are now in Manzanillo. As we are unloading for the night, goodness, there was such a smell of gin when we opened the rear window. Turns out the new bottle of gin leaked and lost two-thirds. Since it is always packed in the box of clothes we will next remove, well, our clothing was soaked. Even the back of Velveeta was soaked. What a mess. I thought Robbie was going to cry through the cleanup, but he just kept murmuring, "Bloody hell."

Our hotel has us sitting at the beach. The waves are gorgeous and within fifteen feet of where I'm sitting. The prices are back to being too high. If we were tourists rather than travelers, we probably would think the price was worth it. As we are travelers, we decide not to make any more side trips along the coast and head directly for Acapulco.

As a side note: While in Acapulco, we intend on investigating whether there are any ships for the three of us to travel to South America in order to avoid driving through Central America and Columbia. I've been worrying about driving through Central America since developing our travel plans. If we must drive through

Central America, we have the route planned out, but it's not our first choice. Besides, dollars are flowing out of our pockets like water.

Driving out of Manzanillo was a complete disaster. Guess who was driving, Robbie. I knew better but he wanted to start us out. It was up and around, back and forth, never-ending. There isn't much point covering it in much detail because neither of us will forget three hours (I'm not exaggerating) of utter and absolute frustration of trying to get out of that city. Seriously, the signs were not correct, the maps were not correct, and typical (or anywhere in the world) the signs just don't display which road goes where. No one seems to travel away from their home city center, or maybe, they just don't want you to leave. Robbie's exasperated look was so sad. I keep navigating and finally just stated, take this turn out of the roundabout and stay on it. We will be out of here. Just don't turn back or off the highway.

That worked and we were again out and off and on to a pleasant ride. And most importantly, headed south. We finally came to nice coastal and rocky scenery. Hairpins turn into the hills. A long drive, we started at eight o'clock in the morning, then three hours just getting out of Manzanillo, and then arrived just around five o'clock in Playa Azul. We have a room with a roof over our heads. The weather is hot, sweltering, and humid. We did get one bad ping on the windshield coming down. That's about it for Playa Azul. Tomorrow, we shall head on down to Acapulco and try and base there for a few days. Change some dollars and find out if we can find a ship that could deliver us to Ecuador. I really don't want to travel through Central America or Columbia.

On May 23, terribly hot. Actually unbearably hot. Driving was not pleasant. And the idle RPM on Velveeta is just going berserk. It is a question of riding the clutch and riding the brake to keep it down. The road was variable along the coastline and quite nice and surprising in certain areas. The truckers were no longer as nice as those up North. The air is once again smoke-laden.

Driving into Acapulco was a complete mess. You have to go through the dives and alleys, then through the center of town. The traffic was crazy. We finally decided to stay at the third hotel we visited. The first two were out of the question mainly due to high prices.

The next day, we changed hotels after a good breakfast. The hotel (Hotel Linda) is a very pleasant one. As soon as we register and pay ahead, workers come in to start working on the hotel. We figured we must be bankrolling the updating of this tiny hotel. It really is lovely and can use some fresh paint and polished marble floors. We smile and just enjoy the pleasant staff and great bar drinks out on the veranda.

While here, Robbie is able to contact a local Nissan shop to have Velveeta serviced. She really needs that gearbox cleaned.

The weather is overcast and humid. It stayed overcast until about four o'clock then the sky cleared. Since we are getting Velveeta serviced and washed up, we should do the same. Laundry to be done costs $25 but so necessary before we depart.

We decide to stay a few more days to make sure Velveeta is in tip-top driving shape. Besides, nothing wrong with exploring the city and its surroundings.

Acapulco

We take a cab to Velveeta. They did a marvelous job on steam cleaning the engine down and the lower chassis. They were able to remove most of the tar. She had an oil change, plus a new oil filter. They also cleaned the air filter rather than use the one we gave them. So we'll save the German air filter for another time. The idle RPM is still high but does not have the roughness or high pitch sound. They cleaned the whole linkage down with a silicon spray, and it seems to be much better. But until we take her out for a run tomorrow for an hour and see if the RPM does increase once more, we really won't know if the problem is solved. Robbie thinks it really is a problem between the roughness of the linkage and just high humidity and slight starvation as to the fuel mixture. Because the exhaust stack is very rich, so she is getting too much fuel.

Saturday, May 27, arrives, and we now have our plan B travel plans before us. We didn't have any luck finding passage to Ecuador. So, we will be driving through Central America and Columbia. This is worrying to me from all that I have read. There is continued fighting in Central America and civil discontent in Columbia. We will need to be very aware of our surroundings and be smart about our travels.

We estimated it will take three days to drive down to the Guatemalan border at Cuidad Hidalgo, which is actually on the Mexican side.

We drive 254 miles to Puerto Escondido. Very pleasant drive. Variable countryside, the coastal route, but didn't see the Pacific Ocean too much. Have seen more of the flame of the forest trees. Very delicate red flowers that set the name.

Flame of the forest

Our drive from Puerto Escondido to Arriaga was another long one of just over 250 miles. The weather condition continues to be extremely hot. Our drive tomorrow will take us into Chiapas, the area supposedly renowned for having lots of bandits. Hopefully, we will miss out and not engage with any bandits!

As it turns out, the drive to Tapachula (through Chiapas) was absolutely amazing. The best roads we have actually traveled in Mexico and the nicest people. I cannot say enough about the beautiful countryside. Good thing we had a beautiful drive because we had little conversation. Our overnight stay in Arriaga was absolutely miserable due to the heat. We thought the overhead fan would be sufficient. But that fan couldn't move the air fast enough to cause even a slight breeze. So we were both exhausted and needed a pleasant and quiet drive.

The rule tonight for finding an acceptable overnight room: It must have air-conditioning. Cost is not a factor tonight!

Tapachula gave us a cool evening's sleep even though I found I had to leave the light on in order to keep the cockroaches at bay. I even slept with my shoes on in order to ensure those little beasts did not get a chance to bed down in them.

We left Tapachula on May 31 to cross the border at Cuidad Hidalgo. I will leave that border crossing to another chapter.

In closing our adventure in Mexico: Everywhere you will find such nice and welcoming people. The drive through the mountains and valleys was more than memorable. I could go again and again. If I should get a *next time*, I think I want to go in a sporty car (no offense Velveeta).

The sighting of the Mexican cowboy and his horse continues to make us smile. I loved the discovery of real ice cream fruit flavors. An easy drive from the east coast to the west coast. Porta Vallarta was much needed, and the small villas were worth every peso. Acapulco was nice but we wouldn't bother to return. Manzanillo was a pain in the neck. Chiapas, we loved the countryside. The drive into the town held magnificent views.

All in all, compared to the guidebooks, Mexico is more expensive than they lead you to believe and plan. Another way of saying it—it is just as expensive as the United States. But oh, so happy to have this adventure at the outset. We feel as though we are now preparing ourselves for more but nothing the same. For the land of Mexico is grand and never the same whether you are north, east, west, or south. It has beauty everywhere, so don't for one-minute stop, looking, and listening.

So as we continue heading south, I realize the necessity of a new mind set. Not everything we expected from our reading is true. I also realize we need to start enjoying what we discover and set aside expectations that we now know are unfounded. I'm starting to become aware of the true meaning of *adventure*. That's a very good discovery after travelling over five thousand miles.

			Travels in Mexico			
Date	Mileage	Number Miles	Location for the Night (or more)	Notes	Days Travelling in Country	Miles Travelled
10-May	147016	233	Hermosillo, Mexico	Drove some extra miles to enjoy the early sights of Mexico.		
11-May	147186	170	Yecura, M	Drove over the beautiful Sierra Madre mountains. I wish I had a Porsche to drive, but Velveeta was great! Road conditions excellent. Incredible hairpin curves, used for passing trucks. Truck drivers very helpful to let you know there was no oncoming traffic.		
12-May	147380	194	Cuantemoc, M	This was a German Minnonite Settlement		
13-May	147452	72	Chihuanoa, M	Lively city		
14-May	147599	147	Jimenez, M	We now start heading west		
15-May	147904	305	Durango, M	Lovely City. Stayed in hotel named Casa Blanc. The drive through the valley was magnificent. Stunning gorges, then rolling hills followed by beautiful farmland.		
16-May	148110	206	Rosario, M	Mountains once again beautiful. Lots of burning occuring. Very smokey. Crossed the topic of Cancer. No site of Pacific yet.		
5/17/- 5/18, 1995	148185	75	Playta Novillero, M	We arrived on the beach of the Pacific. Stayed in hotel Miramuar, right on the beach. Surf is wonderful. Staying for couple of days and take care of some needed laundry.	21 days	3157
5/19/ - 5/20	148426	241	Puerta Vallarta, M	Stay at a condo hotel. Very nice. Manager would like us to stay longer but we really can't afford more than 2 days.		
21-May	148622	196	Manzanillo, M	Lovely beach town considering nothing here.		
22-May	148897	275	Playa Azul, M	Took 3 hours to get out of Manzanillo, felt like going around in circles (which we did several times).		
23-May	149135	238	Acapulco, M	What a dump of a hotel, Hotel Avendia, but too late to keep wandering. Will look for another tomorrow. Velveeta is riving too high. Needs a good engine cleaning.		
5/24/ - 5/27/1995	149167	32	Acapulco, M	Hotel Linda. Close to the bay and very nice. Of course, they are going to start remodeling, no doubt our donation in nightly fee is helping.		
5/28/1995	149421	254	Puerto Escondido, M			
5/29/1995	149677	256	Arriaga, M	This is the state of Chiapis. Beautiful mountains		
5/30/1995	149869	192	Tapachula. Mexico	Overnight before crossing border		
5/31/1995	149940	71	Ciudad Hidalgo, M	Mexcio - Guatemala Border. (I had a obtain a Visa, John didn't. All the papers given us had John's name but not mine. So much for advice at the USA boarder.)		

33

Route Map thru Mexico

Chapter 3

Central America—Unexpected Frontiers

Reflecting on Central America is at times befuddling. Every single country presents a different personality. Yet, you can find yourself in and out of each in no time at all. You find yourself spending more time crossing the frontier (border crossing is now called *frontier*) into the next country. The frontier requires a great deal of patience. Mexico had its interesting aspects with my being asked more than once by the American consul if I was certain I wanted to continue with our travel plans. It would be the first of many times being asked that question and I would always provide the same answer. Yes, my partner and I have planned this trip for over a year, and I expect to travel the distance to Argentina.

From our research and the information gleaned from the guidebooks, we believed we were reasonably knowledgeable of what it would take to drive the distance. We were naively content. I should have known better. As a pilot, Robbie knew the *distance* from the air. His instincts were honed on some very sporty adventures that proved his life could depend on gut reflex. Such as flying three feet above the ground at three hundred mph was okay. I, on the other hand, could not have *known*.

Robbie's life included flying in combat during the insurgence of Myla, fighting and killing in Africa, pirating cargo, searching for oil and whales, or assisting a government official or two for one reason or another. Circumstances such as these are seldom repeated. I have had my share of good times when exploring Egypt, but not like I would find myself experiencing Central America and beyond. And of course, I too had my part in contributing to the daily adventure.

Crossing into Guatemala was a very interesting adventure all on its own. This was all occurring on the Mexican side in Ciudad Hidalgo. They decided that I, as an American citizen, needed a visa. For Robbie, being a British subject, he did not need a visa. And our thought that being married would prove mutually beneficial did not prove true, never. It did prove I was traveling voluntarily with the British gentleman, but I was always asked, "Are you sure you want to continue the travel?" Other than that it was completely reverse to what the consulate informed us in the United States.

In order to obtain a visa, I needed to go to a different building. Robbie was not to join me. I was deposited in a colorful canopied motor-driven tri-ped and transported at the visa issuing station. I only lacked a sun parasol to really look like a lady of a bygone era. Actually, I was quite worried, knowing hardly any Spanish, going at it alone, and back toward the original point of entry. Just me and the tri-ped driver. I was already having visions of not being permitted to return and *fighting loose* to reach Robbie! Yes, I would! Robbie's words of encouragement were to speak in English slowly and clearly in a voice of authority and throw in a few cuss words. If that doesn't work, cry. Latins are pushovers when a woman cries! Great, I should make a scene. No, thank you.

My worries were set aside because a very nice young man at the consulate welcomed me. The gentleman at the consulate was American and very nice about it all. He just kept saying, "Are you sure you want to do this?" I just kept smiling and saying, "Yes." Looking directly into his eyes, I asked, "What is it that you should be telling me because I do want to go?" He had nothing to say. He simply looked at me without any expression and stamped my passport. I was back in the canopied motor-driven tri-ped smiling with a naive belief that future border crossings could not get any worse than this. The tri-ped driver took me directly to where Robbie was waiting. I was back waving my passport and looking absolutely radiant.

Off we drove along CA2 through Tecun Uman and onto an overnight stay in Escuintla. We stayed at Posada, Don Jose. Which so far has not struck us as that marvelous as the guidebooks would have you believe. It really needed freshening up, but it has a lovely structure. The prices look as though they are the same, as in Mexico. The exchange rate officially was 5.65. I stand corrected, $1.00 will convert in the bank to 5.63 quetzals.

The next day, and all of sixty-eight miles later, we cross the border at La Hachadure into El Salvador. Interesting in that, there is a single land bridge at the crossing. That's one way of controlling entry and exit. Just one at a time. Fortunately, we were the only vehicle wanting to cross in either direction.

We continue along highway CA2 and into La Libertad, just south of San Salvador. Our plan is to always stay away from large cities. One more night in El Salvador and we are into Honduras. The drive through the countryside is lovely. You can see just how poverty-stricken the country is but it really does have potential if they could get investors and a government that would give incentives for businesses to come alive and hire its people. The same is true for Guatemala and I suspect we'll see the same through Honduras and Nicaragua. No wonder everyone wants to come to the United States. They just want a chance at the possibilities of life.

From La Union (last night in El Salvador), we head to Sirama (slight backtrack) to catch Pan-Am Highway 1. This will take us to Honduras and the border crossing at El Amatillo.

At this point, we come upon a bit of a surprise. The road has us in the countryside and we see a grass airfield to our left. Then the road turns and we come upon an interesting view. There is a very small bridge and chain, crossing the road with several men sitting alongside the road. Well, this is interesting for a border crossing.

Now, I will admit my action was probably a bit over the top, but I simply could not just sit there quietly in my seat and do nothing. Robbie had stepped out of Velveeta and greeted everyone and stated we were wanting to pass into Honduras as we were driving to South America. This took about ten minutes and Robbie returned. He reported the group of men and boys gathered opposite of the chained gate, simply smiled, and said, "No." I turned to Robbie and said, they must have said more and asked for money. Well? Robbie replied, no. They simply stated no entry.

My temper went up, and I replied, "Well, let's see what they think about this!" I got out of Velveeta, took hold of the hook on the winch, pulling the hook, and attaching it to the chained gate. Then gave out the call to Robbie to *heave-ho*. He did and I stepped back and watched all of their horrified faces start screaming and waving arms in the air as if to say STOP. The guard on the station bit his cigarette in half at the sight. I did find my sense of humor later when Robbie indicated Latins were notoriously bad shots and I would probably have only finished up with the odd flesh wound of the leg. They were apparently screaming, "That redhead has gone mad." They removed the chain. Robbie gave them a salute as we drove through. I, on the other hand, just stared forward and never looked back. This would not be the last adventure in crossing a country's frontier.

Needless to say, we simply barreled through Honduras and didn't want to spend any time there. So we continued our route and two hours later, we were at the Nicaraguan border at Guassale. This can't possibly be that bad! Still, such positive thinking about human nature, and so naïve even at this point!

Now, before we left the good old US of A, we inquired across a number of consulates, travel clubs, and so forth, about visas for ourselves and permits for the vehicle. This exercise turned out to be rather frustrating because no one really seemed to know anything about driving one's own vehicle apart from we would not be allowed to sell it. As it turned out, during the whole journey, we had only a few little problems with our personal visas or passports. It was the vehicle that they really charged for. For the most part, most played by the rules, and were really no trouble. But Honduras and Nicaragua will always have prominent memories when it comes to Central America.

Why Don't We Drive from Portland, Oregon, to Argentina?

So, let me try to calmly share our experience with the border crossing at Guassale, Nicaragua. This experience would set us into totally altering our mindset regarding border crossings and any official we might come upon. Why, because they wield power.

At this border crossing, we came upon the nastiest of fellows of the whole trip. What started out as a $20 bribe payment finished up as $200, "Pay up, or else?" Yes, we did become straightforward with bribery because it worked, and it was expected. Robbie was convinced this one would have wrestled a rhino to the ground, with one hand tied behind his back. The first thing the official proclaimed was the title of the vehicle was not valid, and the chassis number was different from the engine number. Robbie was walking about saying, "This is bloody nonsense!" By this time, we did mental calculations, this fellow was going to make more cash from us than he did all day. Just to emphasize his point, he led both of us to a door that opened into a very large warehouse that was full of very sharp-looking vehicles. He grunted the word, *impounded*, then muttered through recently capped shiny white teeth, "No pay, no leave, jail."

Alarm bells and flashing lights are now filling the brain's sensing devices. His last comment to Robbie was he would not issue a vehicle permit unless we paid a very large fine. *Multa* means a cash on the spot fine. According to Robbie, this is a very popular form of relieving money from some unsuspecting soul who runs afoul of the local officials.

Whilst all this bantering and threatening is going on, I returned to sit in Velveeta. I would cause less trouble, watch over Velveeta, and bargain dollars for local currency. I would always say we would have more in our pockets than theirs. And the money men were always caught off guard by my *female* unassuming nature, but unbridled temerity to *go for it* and always come right out and get a deal on the currency exchange.

Back in "that man's" office, Robbie spoke of how *that man* relaxed in a huge chair, which was large enough to accommodate a five-ton gorilla and still have room to spare. Robbie described how "that man" put his highly polished Kalvin Kline (or whatever the trendy shoe brand name is from Miami) boots on the desk, then tells his sidekick (who is not as nattily dressed) to inform the Gringo that he will not issue a permit. We cannot go forward or return to Honduras unless we pay a fine. Plus, he does not like the British!

That did it! Robbie didn't hold it in and let go of his Spanish. After ten minutes of calling doubts on *that man's* parents and his sister's profession (not to mention his drunken debauching wife sired by a mule with features to match added in some French and Dutch, which are marvelous languages for insults). Of course, the officer could not understand a word, but Robbie was certain he got the meaning. By the time Robbie finished the realization that our *fine* had probably already doubled, if not tripled, hit him right between the eyes. Robbie said he hated to break the news to me knowing I was haggling in order to get a decent rate of exchange. Even Robbie admitted I had become a pro! Actually, I just loved to see their facial reactions when I said no to their extortionate rate of exchange offers.

All these charges for permits, custom inspections, even the fumigation are not paid for in local currency, but the almighty USA greenback. If dollars are not available in your pocket, they are willing to sell you some at an extortionate rate so you can then pay your fine! Seriously! You end up giving it all back to them. They really are well organized in their crooked, but well-managed moneymaking.

Those infamous and misleading guidebooks will inform you that permits will cost no more than $10.00 or so, and under no circumstances should one encourage bribery. They do fail to inform you that if you do not pay you do not pass GO, but go directly to jail and say that your vehicle is impounded along with one's belongings. Obviously, one does not have a choice, unless you are demented.

I will not present the ensuing conversation between Robbie and me but suffice to say, I brought out the dollars. Also, needless to say, Robbie had a turn at receiving my *spirited reaction*. After shelling out $200 of hard-earned greenbacks, we received our papers plus another flash of sparkling white capped teeth. Robbie so loved the thought of sending *that man* spread in little pieces all over the floor! "That man" still had the last laugh though because although he had stamped all the paperwork, he had not signed them! A mile or so down the road, I scrawled an illegible signature on everything, and Robbie yelled out, "Up yours, cocky!"

But there was one item neither of us could change. "That man" gave us thirty days on our visas for Nicaragua, but guess what, he gave Velveeta, seventy-two hours. And as the form read, if not complied with a *heavy fine* of $400 local would be levied. This could roughly be translated as $500 at least at the point of departure, based on the graduated entry fee we had forked out. Funnily enough, we had planned on taking at least five days driving sidetracks to the volcanoes and gorgeous lakes that abound in Nicaragua. So much for that plan.

Exasperated, we tried to check our emotions and simply drive onward to Leon. Thinking back over the day with my pulling the winch out to drag down the chain linked across the border entry into Honduras, then the horrible entry into Nicaragua with *that man* had to cause you to shake your head and find some laughter in this episode. Fingers crossed for some gin and tonic when we arrive in Leon!

Inn in Leon

We were lucky to find a charming inn in Leon. Lovely city, but we could not stay to explore.

Nice bath for Velveeta

    The next day, we were off for Riva, just south of the volcano, surrounded by a lake. Gosh, I really wanted to explore this area, but we needed to get Velveeta to Costa Rica.

    The frontier crossing into Costa Rica at Penas Blancas was a complete joy. Probably because we were so apprehensive, but the border officials were so pleasant. No hassles, everything by the book.

    We continue on Pan-Am Highway 1 then took Highway 3. We managed to make it to Puntarenas and settle for a couple of nights in a lovely hotel. We are back at the Pacific Ocean once again. Wonderful!

    That evening, we enjoy by just relaxing and sitting by the poolside to enjoy dinner. We watch everyone gathered around a TV at the bar, watching the world soccer tournament.

    Finally, a gentle time for planning the next steps into Panama. Unfortunately, the next night, Robbie is caught ill. So ill I had to have a doctor come to call. Robbie lost all his food and fluids. He was running a high temperature and was eventually put on an IV and confined to bed for two days. The doctor discovered it was the shellfish that gave Robbie food poisoning. He recommended we avoid all shellfish in the future. Finally, after four days of normal rest and recuperation, we were back on the road, and I was once again in the driver's seat. Robbie was still sheet white, but at least, coherent and speaking English again. He only spoke Spanish while the fever took control. That in itself was both bothersome and worrying. I couldn't begin to feel as though I was helping and continued to have a nurse stop by once a day.

Finally leaving Puntarenas, I head us back into the country and take Highway 3 back to Highway 1. We travel around San Jose to catch Highway 2 and onward to San Isidro El General, our next stop for the night. Robbie is looking much better as we travel around San Jose. He is still feeling very weak and concerned about his ability to keep moving every day. We'll rest for the night then drive into Panama. When we reach Panama City, we will give ourselves a few restful days (away from shellfish) and that should return Robbie's energy.

Quiet night and much sleep for Robbie at San Isidro. I continue driving us the next day to the border crossing at Paso Canoas, a truly border town since it is located in both Costa Rica and Panama.

The border crossing into Panama was straightforward and had no surprises. We were off and it felt wonderful. We wanted so much to quickly reach Panama City but would find ourselves needing to stay the night in Santiago.

The next day, Robbie was feeling *back in the saddle again.* We continued on Highway 2 and our excitement of arriving at Panama City was growing. After a lengthy conversation of what to do and where to go first, it was decided. I read to us out of the travel guide about the various hotels in the city. We decided we must make our way to the Hilton. There, we would sit at the bar and enjoy a couple of gin and tonics before making our way back to a hotel that overlooked the bay and would give us an opportunity to watch the passing of the ships into and out of the canal. So exciting!

Then it all came before us. The Bridge of Americas and the site of Panama City. As we drove across, I saluted the many flags and looked down at the glorious sight of the Panama Canal. It was fabulous looking across and down where the ships were moving through. This was going to be a fun city to stay in and explore for a few days.

Panama City entry

Once across the bridge, I gave Robbie directions on getting us to the Hilton Hotel. Our driving rules were for me to drive in the city, but Robbie would stay at the wheel. He wanted to be the first to drive over the bridge. I would get my turn later. (Yes, that's right. I'm not leaving this city until I get my turn to drive across this bridge!)

I had a guidebook in hand and a city map to navigate and land us in front of the Hilton. It wasn't our plan to stay at the Hilton, but rather to slide up to the bar and have two ice-cold, well-made gin and tonics. Upon entering, everyone in the lobby just stared at us.

Both of us wearing hats, long pants, and climbing boots. We walked up and set right in the middle of the bar. The bartender asked where we were from and we proudly gave a brief description of our crossing down from the USA, specifically Oregon, and driving to Panama as we were on our way to Argentina. The bartender stated we looked like we just walked out of an Indiana Jones movie. We looked at each other and said, "You probably aren't half wrong!" While enjoying our G and T, I researched the guidebooks and found a hotel on the side of the hill looking out upon the opening of the Panama Canal, Pacific side. That would be our base for exploring Panama City. Lovely!

There are three activities that were on my list before we thought about leaving Panama City. First, find the hotel where President Teddy Roosevelt went in order to organize the effort of the Panama Canal.

Old Panama City

Second, we drive back over the Bridge of Americas, take a few more photographs, and I would have my turn at the wheel for *The Bridge*.

Standing at Bridge of Americas—successful drive over!

Third, we drive up and down along the Panama Canal, getting a good glimpse of the ships and movement through the canal.

Many years ago, I experienced the Suez Canal and was amazed standing at the Suez's edge and almost able to touch the ships as they passed through and slowly made their way from the Mediterranean into the Red Sea. The Panama Canal was very different in size and in the use of the locks. And you could not begin to reach out and touch the ships. But both canals were fascinating and grand.

Our most relaxing adventure though while in the city was just sitting by the poolside and watching the movement of the ships during the day and night. A city constantly with activity day and night.

Robbie made a number of phone calls and found us a ferry to transport the three of us from Panama to Columbia, departing from Colon. We had discussed driving through the Darin Straights, but another time.

Looking back at the ships in wait for their proper hours of passage through the Panama Canal brought a wonder of thought about the efforts needed to align the international commerce on a single thread of timing. At night, ships move from the Atlantic to the Pacific, during the day, the flow is Pacific out into the Atlantic. The tonnage moves. Panama City controls the traffic through the canal. And it is very organized. Fascinating to watch.

Looking out to the old city

    I searched for evidence of those days of Teddy Roosevelt in Panama City. Looking upon the old city, its face barely expresses a hint of the dramatic revolution of events that occurred at the turn of the twentieth century. The digging of the canal, the political drama, the human loss, and a step forward for international commerce. Undeniably, the step forward never seemed to include the people of Panama.

Tivoli Hotel

    Taking a glimpse and stepping into the old hotels that once were lodgings for those visionaries showed a silent worn glory. Those that let me enter were surprised at my interest. They indicated that no one ever just walked in unless they were with a tour group. Standing in the grand lobby, walking up the grand staircases, and looking down. You knew. The grandeur was still there, you only needed to look beyond the absence of furnishings. Its design and cleanliness silently spoke of times that no longer lingered in anyone's memory.

    Now, as we step into another turn of the century, Panama City will experience yet another change. Teddy's big stick departs, and the Panamanians gain control of the canal. Hopefully, their vision will continue this incredible flow of commerce on into the next century. The city, in itself, is grand and fascinating. I'm certain those two characteristics will remain and grow along with the city into the twenty-first century.

Thank you for a fabulous visit. The frustrations of the past border crossings are now long forgotten because of the pleasure of Panama City.

		Guatemala / El Salvador / Honduras / Nicaragua / Costa Rica / Panama				
Date	Mileage	Number Miles	Location for the Night (or more)	Notes	Days Travelling in Country	Miles Travelled
5/31/1995	149943	3	Tecun Uman, Guatemala	First town after the border crossing, taking route Pacific Highway, CA2		
6/1/1995	150020	80	Escuintla, G	Roads are terrible. Took 6 hours to drive 60 miles. Continuing on CA2		
6/2/2020	150088	68	Border Crossing, La Hachadure, El Salvador	Bridge crossing interesting (only 1 lane, and fortunately we were the only ones crossing in either direction). No difficulties.		
6/2/1995	150163	75	La Libertad, El Sal	Stayed south of San Salvador, continuing on CA2		
6/3 - 6/4/1995	150285	122	La Union, El Sal	Stayed in Hotel San Francsico, Need to stay extra day for laundry and prepare for trip across Honduras in a day then into Honduras. Will drive back to Sirama to pickup Pan Am route 1 for border crossing.		
6/5/1995	150320	35	Border Crossing, El Amatillo, Honduras	Border Crossing at El Amatillo - small bridge with chain across. Refused to let us pass until I hooked winch on chain and we began to pull down chain. They changed their mind. Barrelled thru Honduras.		
6/5/1995	150393	73	Border Crossing, Guassale, Nicaragua	Border Crossing Guassale, Nicaragua. Approximately a 2 hour drive. After horrible time at border crossing. Threatened to have Velveeta taken and placed under house arrest in a barn,		
6/5/1995	150462	69	Leon, Nicaragua	Long, Busy & Frustrating day. Left El Salvador, crossed into Honduras, drove thru Honduras, and crossed border into Nicaragua. God, what a day. Now for a quiet evening of rest in Leon.		
6/6/1995	150579	117	Riva, Nicaragua	Last day in Nicaragua. The drive was beautiful. Seeing the volcano surrounded by a lake so prestine. It is said to have white dolphins living in the lake. Can't stay to find out!	19 days	1460
6/7/1995	150600	21	Penas Blancas, Nicaragua Border	Arrive in Costa Rica the morning of 6/7. Did not want to spend another day in Nicaragua. Know it is		
6/7/- 6/12, 1995	150731	131	Puntarenas, Costa Rica	Unfortunately, Robbie ate shell fish the 2nd night and became very sick. Ran very high fever. Doctor was sent for and put Robbie on IV. Really took a toll on him. Very weak.		
6/13/1995	150893	162	San Isdora, CR	Beautiful tropical forests. Rained all day.		
6/14/1995	151,030	137	Border Crossing, Paso Canoas, Costa Rica & Panama	This town is in both Costa Rica and Panama.		
6/14/1995	151173	143	Santiago, Panama			
6/15/- 6/18, 1995	151341	168	Panama City, Panama	incredible city. Driving over the Bridge of Americas was fabulous. We first went to the Hilton Hotel. Walked up looking like we came out of an Indiana Jones movie. Went to the bar and had gin and tonics. Talked about the adventure, then left to find a recommended hotel - Hotel Montreal. Up on the side of a hill overlooking the bay. Pool and so relaxing. We deserve this splurge. Drove into old town and found the hotel that President Teddy		
19-Jun	151397	56	Colon, Panama	Lovely drive to Colon. Lots of bill boards advertising all sorts of goods for sale. Very tropical.		

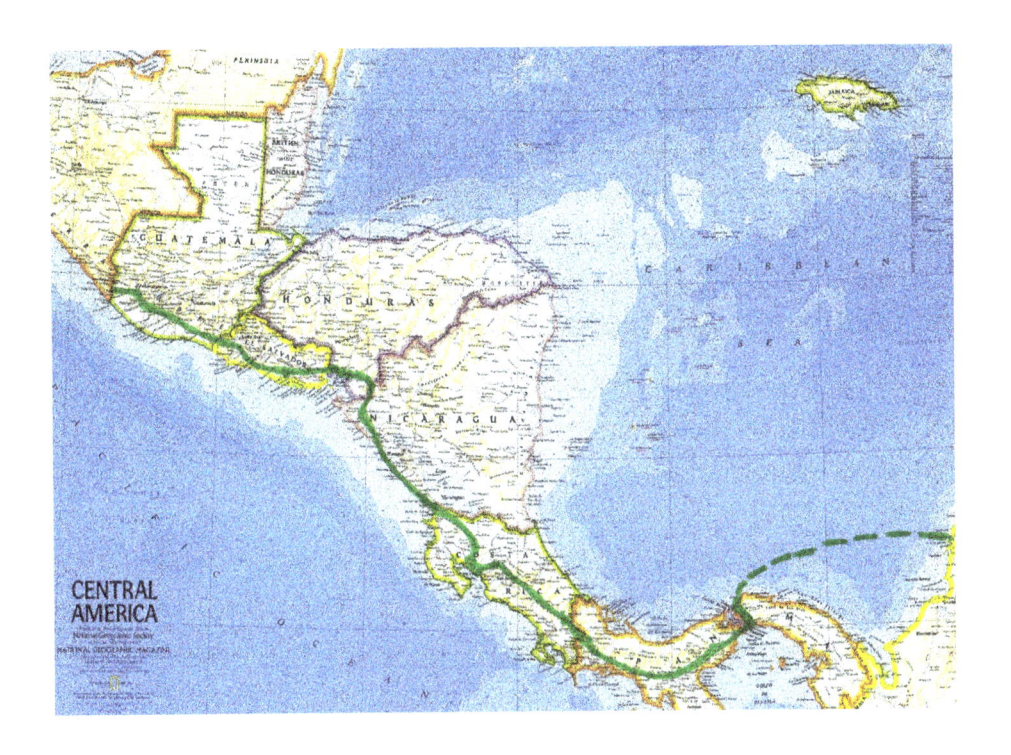

Route Map thru Central America

Chapter 4

Passage into South America

Headed to Columbia, saying goodbye to Panama

What a glorious feeling leaning over the railing of this lovely ship and stretching out toward land, the city of Colon. I didn't want to let go of our memorable travels to Panama. The drive from Panama City up to Colon was also fascinating. Lots of trucks and more advertising billboards than anywhere else in the world. Anything you might want is probably somewhere in Colon!

Now, we set sail. The cool breeze is blowing our scent back to land, and we were setting out again on yet a new path of our journey through the Americas.

Embarking out into the Atlantic Ocean and sailing to Colombia brought feelings of excitement for this next episode. Thus far, we have traveled just over 6,900 miles. When we had the ocean, it was the Pacific. It gave us an oceanic view of the constantly changing blue Pacific. Now the craggy grey of the Atlantic would become our floating carpet south.

Yes, indeed, we were floating south for entry into Columbia. The thousands of miles we have driven have helped to reduce the anxious feelings of this next frontier entry. I just keep thinking about the press and constant barge of stories that always places Columbia, and thus all of its people, in an anxious area of the mind. It isn't fair, it isn't even a true picture of the land and its people, but whether we want to call it bad press or manipulation of the facts for Government disinformation, the fact is the mind is already in a position and responding in a very spontaneous fashion that even beguiles the most avid liberal! What does all this mean? I was a wreck! If we were lucky, we wouldn't be shot at the border!

Between moments of anxiety attacks and excitement of the adventure, I often did not speak. For now, on to Colombia, then we continue our drive on the Pan-American Highway into the Southern Hemisphere.

The soft yellow of Robbie's cravat showed he had lost his *moon man* color. A nice bronze tan looked healthy with his blue eyes. He was always warmly welcomed, except in Nicaragua. Difficult to say whether it really was his British heritage or his frightfully surprising Spanish!

As we walked into the ship's large greeting area, we could see the very proud and youthful crew. Dinner would definitely be a treat, followed by an evening of entertainment and dancing. I had no idea a ferry could be such a delightful ship. We drove Velveeta aboard the ferry and made sure she was locked and left her for the night for a well-earned rest. We'll be back in the morning.

We walked beneath the waterline and found our cabin. Even though Robbie's nose had already found the bar and he was sheepishly smiling and wanting a Ginny Winny, I needed to check out the cabin. We bumped against the small hallway toward our small, tiny cabin. Two bunk beds, but very clean quarters, and a very clean shower and toilet. Definitely, no need for the checklist tonight. We were starting out on the right wave.

Robbie, taller than the entrance door, smiled with approval upon entry into a whiter than white cabin. Unsure whether to stand straight or just move forward in one motion, we both fell into the small bottom bunk bed and just laughed. Tiny is still inviting, and you could feel the *goodness* about cleanliness and freshness that can be touched with a lingering softness to the fingers and palm of the hand. We were full of giggles and playful times. Our one-week much-needed rest in Panama City renewed our strength. This, our maiden voyage (anything but maidens on this ship!) presented another new chapter in our adventure and we felt the energy. Love to revitalize the energy!

A few minutes before our designated dining hour, we made our way through a score of people looking full of life and fun. Most of whom were at the age from midthirties to sixty-something. Just right!

An absolutely perfect four-course dinner, full of wine, and a pleasant threesome conversation accented with an American smile. Robbie and a very lovely Colombian couple hit it off immediately. They were charming in every way, and Robbie, as usual, so well received. You don't have to thoroughly understand a language to capture the polite, yet warm, manner. And speaking of the dinner, salad, soup, entrée, and dessert, with lovely wine from Colombia. Just the beginning.

After dinner, we made our way through the various tables, a nice club effect. In sight, at the back of the room was a huge bar and lots of waitresses quickly serving guests. Everyone was wanting to dance, but they all appeared to be like an older set. I couldn't imagine what this would be like. Robbie smiled, "Oh, you just wait and see. South Americans are going to surprise you."

After the fun of the Vegas-style show, the real entertainment began. The dancing of the guests was a celebration. It was necessary to sit and watch for just a little while. The opening show brought a band and singers. If I didn't know better, I would think we were going out on a seven-day cruise, rather than an overnight ferry trip. The music had a great rhythm and movement. Not the *Latin* beat that comes out of Miami or Mexico, but a full movement from the heart. The first glimpse came from the performers, but then, when I saw it on the faces of the passengers as they took their turn on the floor, well, I knew I was stepping inside of an unknown world. Colombia was free.

Our turn on the dance floor brought us laughs and our dinner partners brought me the dance movement. What a wonderful movement. It is a movement that can be seen and felt throughout all of South America. The rhythm of the body rolls from the heart down to the hip, from side to side, gently, and with your partner. The rhythm continued for hours into the morning. What a wonderful way to step into South America and push aside the nervous tension. The music—fabulous! The people warm and gracious.

The morning sun brought us in full speed along the coast, and by midmorning, the Cartagena cityscape was visible. Now it was just a matter of time before the next episode of our journey into South America began. Neither of us knew what to expect. Although Robbie had lived in South America for twelve years his knowledge was from the air. As a flight captain, he knew the terrain, but not from the ground.

We went forward having read our books on Colombia and knowing where not to go. We spent many hours reviewing the route. Robbie had laid the initial course, but I reviewed every map against guidebooks and financial bank listings. Together, we studied each route to be certain our needs would be met as we headed from Cartagena. It needed to be planned even with the current events in mind. We knew that we just needed to stay in the small cities and just keep moving. But now, the time came for us to join Velveeta and be ready to disembark the ferry, our QE2, into South America.

Departed and Sailed evening of 6/19 thru morning 6/20/1995	292 nautical miles from Colon to Cartagena	336	Travel from Colon to Cartagena via Atlantic Ocean	International Waters

Chapter 5

Entry into South America

Entry into Colombia was interesting (to say the least) in a morbid and pathetic way. Immigration for the two of us was efficiently and pleasantly carried out, and *free*. Velveeta was issued an entry permit as easy as falling off a log, or whatever one should fall off! We were happy and still feeling exuberant from our water carpet ride into South America.

As it turned out, the immigration proceedings were too easy. As we drove forward, we were confronted by a man dressed all in black. He was obviously modeling himself as an ex-Gestapo officer from the Third Reich. Why do all the bad guys wear black? I think the only good guy I remember who wore black was Hop-along Cassidy, but his guns had white-handled grips.

In front of us, in line, was an old beat-up VW van at least twenty years of age. Loaded to the gunnels with all sorts of junk that was being unloaded by a much perspiring hippy who kept playing a fiddle. This is when my apologies extend to the true country musicians. This guy could not stay out of jail with his playing, but he never stopped. Not sure if it was for his benefit to quiet his nerves or distract the officer who was intent on rummaging through all the junk. There was another black-clad officer, wearing a sidearm that would have made an elephant gun look like a peashooter. He was overseeing the inspection and periodically glaring at the hippy fiddler. Maybe playing that fiddle was the right thing to do because his van did eventually get reloaded and *out of there*.

Our black-clad inspector struts up and requests us to unload—*now*. Robbie, politely says in Spanish, "No way, cocky." That got us nowhere, except Robbie started unloading. Fortunately, our tailgate had one of those automatic rear windows that when the right button is pressed it disappears into the bowels of the door, and then you can fold the whole thing down. The inspector is gazing his eye forward into the back of Velveeta and he stays glued to Robbie's side. The first thing the inspector sees once the back gate is lowered is our porta-potty. This item never ceases to amaze and baffle all *officials* that inspect Velveeta. He stares at the potty, then at Robbie as if to ask, "Why?"

I really think that every officer seeing this tiny potty is trying to figure out a way of making it their very own. We discovered while traveling outside of the ole US of A, toilet seats are extremely rare in every inn. There is an expression Robbie always uses: "I must meditate on this fact one day! When I find all of the toilet seats that are missing from every inn and hotel!"

The inspector removed every item and every container was opened. The clothes were handled as though he knew he would find contraband. He became so frustrated at finding nothing. He came for me while I sat in Velveeta and he demanded that I pull out everything in my purse. Stepping out of Velveeta, I looked directly into his eyes as the purse remains on the floor and said, "You want it, you do it." Robbie spoke out, "Now, darling, help the nice officer."

Well, I just pointed to the purse and he promptly picked it up turned it upside down dumping everything into the seat. He went through it all and then finished by giving me a perplexed look of disgust then turned around and returned to Robbie's side. I turned to look at the woman from the American consulate, standing and just watching the spectacle while she stood inside the inspection booth. She looked back at me and quietly mouthed the words, "I'm so sorry." I tilted my head to one side and mouthed the word, "Really?"

After a bit more prodding around by the inspector, more out of curiosity than anything else, he finished by warning us not to drive at night, too many cows on the road. He did not mention people on the road. Obviously, the cows have a priority status. We were stamped as OKAY in our passports, provided paper of inspection, and waved on. Those three hours of inspection were to stay with us for some time as we attempted to leave Cartagena. Robbie was driving because he felt it better for him to be at the wheel leaving the inspection area. He didn't want anything else to come into the inspector's mind by seeing a woman take over. We both felt absolutely exhausted. Navigating through the city to find our road out took a bit more time but that was at least a distraction that was acceptable if not providing some degree of relief. We have got to move on.

Once out of the city, we started to feel a little more comfortable with the sight of the country ahead of us. The road was straight and the countryside quiet and peaceful. Robbie found a spot to pull over. I took the wheel in order to give Robbie some time to recompose himself after a thoroughly frustrating and exasperating entry inspection. Little could we have imagined of what would be in store for us as we continued our travels.

Back at the wheel, but now in South America, and I'm thrilled to be driving us south. As mentioned, the countryside is tranquil with the breeze blowing through the fields and the trees line the highway. To the front of us is a sporty red convertible speeding along. I'm thinking, how fun is this? We are now out and about in Columbia.

Then as we look ahead, coming into view is a booth with two officers standing to the side of the road and waving down the sporty red car. Robbie and I look at each other and I state the obvious, why isn't the little red car braking? The sporty red car seemed to just step on the gas, and zoom, gone. Those two officers watched the car zoom by, then looked back at us, and began walking out into the middle of the road, obviously making a statement to us that we needed to stop. I was not going to attempt to speed up and drive around them. Nope, stopping is going to happen. Robbie is now saying, "Bloody hell."

I brought us to a complete stop right in front of the officer. He seemed to be somewhat confused in looking back and forth at the two of us. The officer does not walk to the driver's side, but rather, marches up to Robbie. Of course, that immediately puts me in my place. The officer greets Robbie and then starts his banter. Robbie looks over at me and states the officer wants to examine our papers. The officer states this is a standard routine checkpoint that we'll continually meet along our route. We gather paperwork in the glove compartment and my purse. Robbie carefully organizes the individual papers along with the entry papers provided by the port of entry in Cartagena. The officer looks as if he is carefully reading through each, then turned to Robbie stating, "Insurance papers."

I looked directly at the officer, then at Robbie, and repeated the words, "Insurance papers?" My heart skipped a beat because I knew we had no such paper. No one at any border either issued or asked for insurance papers. Not even discussed by Sanborn prior to entry into Mexico. None of the guidebooks.

No surprise when the officer, smiling at Robbie, pulls out his official looking booklet. He thumbs through the pages and promptly shows Robbie a paragraph explaining that all vehicles, not just foreign ones, must have valid insurance papers. It then went on to list the various amounts of fines for outdated insurance, insufficient coverage, or no insurance. The fine is $500, Robbie turned bright red, then sheet white, and I stated, "Robbie, we have not been to an ATM since arriving, and we do not have $500." At this point, the officer signals that Robbie is to follow him. I don't really think the officer understood what I said, but he definitely understood the emotion in my voice.

Previously, I mentioned the art and fun in bargaining through Central America to get a good currency exchange. The exchange agent starts high, and you start low. And so it goes until you meet in the middle and agree. With this particular officer, he was not interested in pleasantries, and he definitely wasn't going to entertain a *high-low* conversation. Robbie followed the officer to a small building, and I followed them with Velveeta. At one point, Robbie comes out with arms flinging, and comes inside Velveeta, partially screaming, "He is going to throw me in jail if we don't come up with the money."

I turned, pulled out my wallet, and emptied everything from the purse. This is it, Robbie, handing him the remaining dollars. All we have is what amounts to $50. Robbie took the money and slowly walked back to the building. What seemed to take an hour was probably only ten minutes, and Robbie at last returned. Robbie stepped up into Velveeta with a look of disgust, showing the insurance papers, and then placed them in the glove compartment. The officer was very clear in stating that papers would be asked for as we traveled through Columbia. The officer also accepted the $50 and the story of our entry that very morning and no opportunity to obtain any additional cash. The officer was not interested in keeping Robbie there nor detaining Velveeta. Just move on.

We needed to get the hell out of there. As I pulled out the drive and looked back there, the officer stood, giving us a very smart salute and a broad smile. I wanted to turn around, jump out of Velveeta, and slap him for all of the aggravation. Robbie looked like he could use at least six gin and tonics. But no, let's just follow our lessons learned from Central America, and now, add two more from our welcome to Columbia. God, what a beginning.

Just in case you want to flip ahead through the pages and find where we are stopped for insurance papers, I'll save you the trouble. No one ever stopped us from that point on through Columbia (or any country) to ask for anything—NEVER! Not even at the border crossings.

What an unforgettable welcome to Columbia and entry into South America. Adventure on!

Chapter 6

Columbia Hospitality Plus

Our first night in Columbia was in the lovely town of Sincelejo. Only 118 miles from Cartagena, but we needed to stop for the night, and refresh ourselves with a spoonful of sanity and some smiling faces. As it turns out, Sincelejo has a lovely town square and the most beautiful cathedral, St. Francis of Assisi. The town square introduces us to a hotel and restaurant that gives us a chance to sit outside and enjoy the pleasantry of the town's people. Good night's rest and we'll be good as gold in the morning.

In the morning, we have a pleasant breakfast, and then we are off. Today is the first day of summer, June 21. We'll get to keep summer until we reach the equator in Ecuador, then we'll skip autumn, and jump into winter. Zoom!

Leaving Sincelejo and continuing south, we are continuously going up. Our route (Highway 25) will take us to Santa Rosa de Osos today. It will be a long day for close to 250 miles.

Now, we were beyond the distractions of entry inspectors and roadside police stalls giving us a chance to discover the emerald green beauty of Columbia.

Emerald green Columbia

The sights are almost unending in emerald-green. The Cordillera (the mountain range extending from North America through South America and includes the Andes) is deceiving, only the occasional glimpse through the trees allows you to know you are on top of the world. Driving through the Colombian mountains is dramatic, maybe because of the absence of wide, sweeping vistas that are as deep as they are wide. The lush trees fill the view with layered blankets of foliage. Hours pass as do the thoughts of altitude. Only an occasional sign provides the reminder—4,800 meters. The mine is briefly fooled until, oops! Meters, not feet! We are not at 4,800 feet, but rather, about 15,800 feet. Wow, that happened so quickly, and without noticing the height. A gradual lift up to the tops of the trees.

At 15,800 feet, we are constantly covered by trees. Cannot tell how much farther up the mountains actually rise. Every now again, the crest of a turn brings the vistas that would indicate up was more, but no real clue of how high we were at this point.

Surprisingly, the traffic is heavy with 16 wheelers, and it is amazing to see so many. These truckers really move up a steep grade. They do not mess around.

At one point, the traffic congestion is so heavy, and the movement is slower than a slug. We are starting to worry thinking, *Bloody Hell, another officer station stop*. Then we see the reason.

Now, remember we have been driving up the mountains of endless greenery. We then see lots of pull-off points on the side where trucks are being (wait for it…) washed. Seriously, we are stunned, and also elated that our first fear of police is not to come true.

This is an amazing sight. Water is being piped down from what must be a lake still above us at the top of the mountain. Young boys are running around these 16 wheelers, scurrying to finish each rig and bring in another. The activity is frenetic. You can see the water being brought down through a series of spouts from high above. Down the spouts come, making a bend here and there, and ever-flowing. Water is, at times, splashing out. The spouts catch onto the treetops, come high across the road, and then are held at bay. The ingenuity of the spouts is clever indeed. Sprays of water are immense as the boys aim the spouts at the rig's top, sides, and wheels. Water rushes through the spouts and cleanse the trucks with a great force of gravity.

This activity goes on for miles, filled with 16 wheelers, regular trucks and cars abound in the turnouts. The water-soaked young boys are non-stop action. Their legs are skinny from constant movement, their faces are full of smiles, and yet, they are focused on business. The more trucks they stop and clean, the more money they earn. This is serious business.

The trucks are spotless when they roll out. A young boy waves his arms above his head calling out, "Wash, two pesos," and stops the traffic, thus assisting the driver to return to the road. He also waves down and pulls in another customer.

We pull over and stop to give Velveeta a treat of soap and water. Velveeta is cooled and cleansed and we are back on the road again in no time. All three of us are refreshed and full of energy. It was exhilarating!

We are laughing and cheering all the way up the mountain. Don't know if it is the coldness of the mountain lake water or the spirit of youth. Doesn't matter, it works together.

It's time to find a place to stay for the night. I'm driving up a very steep incline into a mountainside town, Santa Rosa de Osos. We find ourselves behind a funeral procession. Yes, a funeral procession. In front of us is a crowd filling the street and the procession is moving so very slowly. The pole bearers are steadfastly bearing the weight of the casket, heads down, and slowly making it up the street. It is so slow that the first gear just doesn't cut it. I pull over and we wait. Robbie is shaking his head, arms slightly flinging about, and bantering something about, "In the entire world, how can these situations always happen. It can't be possible."

I just look across and smile. He mumbles and rambles and we eventually find our way to the town center. There in sight, at the town square, is the hotel, and across the street, the town café. That is the town center design that exists everywhere we travel. We are grateful because that design makes our travels so much easier. Driving thousands of miles does not mean you call ahead every day and make reservations in the next town's Marriott or Motel 6. Those establishments don't exist on our route south. Every hotel and inn is different, inside and out. But we learned that one must be particular about the accommodations. You can't just register and get a key to the room. You must first inspect the room they want to *rent* to you for the night.

On more than one occasion, we have not accepted the room or just left the establishment and found another. We now come with a well worked-out checklist in selecting accommodations. Regardless of whether the hotel is in the town center or its outskirts. Do not take a room without first inspecting and going through the entire checklist.

We have a fresh Thursday morning, June 22. We are getting closer to the equator and the time of year will soon be upside down to us. Summer will become winter very soon. Robbie hardly had enough hot water for the morning shave, but he didn't complain. I, on the other hand, find the toes just can't get used to cold tile floors. I throw blankets and towels down as a pathway. A sense of humor must be on hand at all times. We know this is the basics of life in every town so make the best of it and find new adventures for you. You simply look outside of yourself and let yourself see and listen to everything and one around you.

We get a quick breakfast of coffee and toast. We are then off. Velveeta, still looking freshly clean after yesterday's mountain spray.

After driving through the mountains for a couple of days, we finally find ourselves traveling down the mountain. Next city—Medellin. Its country driving and Robbie is in the driver's seat. This downhill ride gives us a view for definite inspection. It is straight down. I look over through the driver's window and it is an amazing sight. Straight down.

We observe the heavy traffic in the early morning, but then realize it is during the workweek. Obviously, Medellin is the center of work. As we continue to travel down the very steep and endless winding road, Robbie starts to talk to me, which is unusual. Remember, this is a Brit, and typically, while driving, we have a great deal of silence in order to keep focus on the road and its surroundings. Always necessary. But, in a quiet and steadfast voice, Robbie very simply says, "Darling, the brakes don't seem to be working."

I am always a kidder of heart and slightly non-believing, and reply, "What, oh, come on!"

Robbie turns to me and states very clearly, "No, I'm very serious. I was not going to say anything, but for the last twenty minutes, the brakes aren't improving."

Suddenly, from a sheer instinct of terror, I discover the meaning of the *white-knuckle* effect. I have both hands bracing on Velveeta's front handle. At that point, even though we are about to drive into town, it was good that Robbie was behind the wheel for this surprise experience. I'm not sure I would have been either calm or focused. Robbie was able to downshift into second gear, but even at this steep grade, Velveeta was moving too swiftly.

Curve after curve, Robbie was able to keep us from moving too close to the edge of the road. I wanted to scream but didn't because I knew I would cause Robbie to take his mind and eyes off the road. I thought I would just close my eyes and let it pass but couldn't. At times when coming upon slow-moving cars, Robbie would narrowly pass them, with a wave outside to oncoming cars. He was attempting to signal we had a problem, so get out of the way. Thankfully, everyone seemed to sense our difficulty and the urgency of getting out of our way.

It was terrifying. I really thought the next corner might find us off the mountain. It was straight down, and no soft landing could be seen as I look across the road. There was no place to go on my side of the road, it was the mountain wall. Finally, the grade of the mountain started to change, and we could feel Velveeta starting to slow her roll. Not as slow as I wanted, but we were slowing down. In the distance, we saw a street entrance that just might work to allow Robbie to totally downshift to first gear and stop. We made the turn, ahead of oncoming traffic, and came to a complete stop. I was shaking and still hanging on for dear life, looking over to see Robbie. He appeared totally unflappable. I asked if he was okay. His only remark was, "I need to find a mechanic, and now."

At that moment, I really believe he stepped back in time as a pilot, getting ready to hop out of the plane. He instinctively knew that his Velveeta needed to be repaired in order for us to keep going south.

Robbie stepped out and down from Velveeta and went into the street talking with everyone passing by. As I looked upon the conversations, I could see each person was concerned and gave advice. He stopped an older fellow, who had a vegetable cart close by, and explained our situation. This total stranger walks out into the street and pulls a car over. I thought the old fellow might have known the driver and maybe it was a taxi.

Come to find out later, Robbie was shocked to learn it wasn't a taxi nor someone the fellow knew. It was just a good citizen, who stopped and drove thirty minutes out of his way to take Robbie to three different shops to find a mechanic that would return Robbie to

Velveeta and me. I, on the other hand, stayed back with Velveeta. I was certain I might be found and carted off by the police. After three hours of waiting, I feared the worse. Throughout the wait, the very nice old fellow with a vegetable cart stayed close by and watched over me. He kept attempting to assure me of Robbie's return anytime. He even gave me bananas to eat while waiting. Really nice, and absolutely a lovely gentleman.

Robbie finally returned with a mechanic who was able to get us rolling again and pointed us into a direction that was a mechanic alley. An area of the city that was filled with mechanics of every variety. It was like an open field of one shop after another. Each specializing in something different, or the same. We rolled in. Robbie strolled through and found the right specialist. Two more hours and $28 (twenty-five thousand pesos), later, we were on our way. Really, $28, a warm smile, and a handshake. How could we have been so lucky?

During the two-hour period, I scanned through my handy ATM directory (listings for all South American countries) and set out to find a local ATM and withdraw some cash. At this point, we had NO money. I'm already thinking ahead and wondering how I can talk to the cabby and make my way to an ATM. This experience was also harrowing in that I kept thinking my Spanish would cause me to end up in some location so distant from where I started, I would never see Robbie again.

We flagged down a taxi, Robbie provides the driver with the basic information. I let him know my need to make it to a bank, spoken in terrible Spanish, but the driver just smiled and nodded yes. I provided the address and off we went. Looking back, Robbie is standing there, giving me a big smile and wave. Well, here I go. In fifteen minutes, we arrived at the bank, and I stepped out and walked up to the ATM as I would anywhere in Portland. I did find myself instinctively looking back to make sure the taxi had not left me behind. No problem, the taxi is just waiting. I was able to read through the ATM instructions, get what I thought would be enough cash for the week, and walked back. My gosh, I could hardly breathe, but the taxi driver was there, and he was so very nice. Never asking anything. Just smiling at me through the rearview mirror.

Alas, I returned with cash in hand at the right spot at the mechanic's row. The taxi driver was so very helpful (a prevalent characteristic in Columbia and all South American towns and cities). I gave him the fare, but also a very nice tip. He warmly thanked me, and I returned the smile and said goodbye.

The look on Robbie's face when I stepped out of the taxi was priceless. As though he thought I had conquered the world. I (and Robbie) was always very amused at my unabashed, yet intrepid, wanderings to find a bank with an ATM. I would be excited to see the bank's appearance, and yet hold my breath until the cash actually appeared. My many conversations with my Citibank personal banker were to make sure they never turned off my card. I chose Citibank because they had always been available throughout my travels in Europe. And the Citibank travel guide did not disappoint me.

Citibank had made relations with a wide-range banking community. I would be able to find a Citibank or their affiliate all through Mexico, Central America, and South America. My directions to my Citibank personal banker was to not let anything happen to my account. I didn't want to find us stranded in South America with no available funds. So far, so good. Our experience thus far demonstrated credit cards were not prevalently accepted, cash was always wanted.

At the end of the day, we were relieved to be on the road again. It was very late and time to find a place to stay overnight. Robbie was behind the wheel because the directions provided by the mechanic were very clear and easy. I know, it could have gone very bad considering our experience with Robbie driving in the cities. At this point, we were very tired and wanted to trust those who provided the directions. As it turned out, it really was easy, even for Robbie. Well-marked highway.

Driving out of the city, I found myself remarking about the billboards. There were so many advertising places to stay. I finally had Robbie pull over and stop, "Look at these signs because they are so romantic looking."

Robbie burst out laughing. Something told me I must have misinterpreted these signs but didn't have a clue. Looking at him smiling he leaned toward me saying, "Darling, the culture here encourages afternoon delights." Excuse me, what? I'm not following. Again, he smiles and says, "Afternoon delights, you know. They are pussy pens."

I found myself more confused than educated. What kind of a pen? "Darling, a place to take your lover for an afternoon or evening but sheltered from view." They are very private and very expensive.

Immediately, I was captivated by the idea. Are we talking about a cat house? With a look of disdain, Robbie replied, "No, you bring your own!"

Probably sounding like a schoolgirl, I said, "I must see this."

I want to go to one. Robbie was laughing and yet not, "Are you sure?"

My reply was in astonishment, "Of course, why not?"

So, we followed the signs and found ourselves driving up a slight hill toward a very secure private gate and a security guard booth. Robbie indicated to the guard we were interested in the night. It was all the guard could do to keep a straight face while looking at Velveeta and the two vagabonds. The guard replied, "How many hours?"

I was about to fall on the floor, and Robbie, very calm, replies, "It's for the night."

The guard said, "The charge is going to be $500."

Of course, I'm now about to faint, but somehow, Robbie excused us. Now, at this point, you know we need to turn around. Robbie, with his ever-unshakeable manner, asks, "Could we just pass through and exit?"

Again, the guard smiles and simply replies, "Sure."

So I had my first and last view from the outside of a *pussy pen*. I wasn't sure if I could say it, but I did. Now, at this point, everyone asks, what could you see? My reply, nothing. Absolutely nothing. There wasn't one car parked outside. The garage doors were all closed. No windows could be seen from the courtside. The brick building was laid out beautifully and with absolute respect for privacy.

We circled the complex and exited. Then somehow found our way to a normal everyday hotel in the town center of Caldas. And of course, a lovely town center restaurant. We enjoyed dinner outside with crowds of people about.

We sat back and smiled, laughing at the day's adventures and delights of the day. The harrowing white-knuckle event, strangers to the rescue, discovery of the mechanic's alley, solo outing to an ATM, quick outdoor introduction of pussy pen, and back in the seat again to a familiar town center design. All in just one day.

After the exciting day and lovely dinner, we return to the hotel and looked over the maps. We decide to spend two nights in Caldas and double-check our route on Highway 25. We only have a few more days until we reach our border crossing into Ecuador. And we want to make sure we understand the choices for our overnight stay and ATM locations. Plus, clean clothes would be a good idea at this point.

On June 24, we head out of Caldas to the next destination—Cartago. It's not a long drive, just 135 miles. We did have a couple of interesting episodes on our journey that gave us pause, but fortunately, Robbie knew how to instinctively handle both.

As Robbie drove us through the lush countryside (trees in abundance on both sides of the road), I pointed out men walking toward us on Robbie's side. They were all wearing black T-shirts with rifles over the shoulder. Robbie simply stated, "Pull down your hat, and keep looking forward."

As we drove up, they started stepping into the woods. Robbie gave a quick tap on the brim of his hat. They didn't really look up at him. On the back of their shirt was the word FARC. It was the Columbian opposition militia, and luckily, they wanted nothing to do with us. I had an immediate pit in my stomach, but Robbie just kept us rolling along at the same speed, and never looking back.

Later in the day, we exchanged seating. I was once again behind the wheel as it was afternoon, and we would soon be coming upon our next town (Cartago) for the night. Still with tree-lined highway and no traffic. About a half-mile up the road, Robbie pointed out a military post manned by a single guard. As we came closer, the guard stood still at the opening, carrying a rifle over his shoulder, looking directly at where he expected we would stop. Robbie looked across at me and said, "Pull down your hat, just drive, do not slow down, do not look at the guard, and do not look back in the rearview mirror. Just keep on going no matter what. No need to speed up. Stay consistent."

I followed the instructions to a T. Robbie said the young man looked astonished and never pulled his rifle. Just looked out following us and his face looked amazed. I'm sure no one ever just drove on through, but memory also reminded me of that sporty little red car on our first day in Columbia. Just keep moving and never stop seems to have become our motto as well.

Cartago was another quiet town. We found ourselves with a pleasant overnight stay and a restaurant for dinner.

Why Don't We Drive from Portland, Oregon, to Argentina?

This is our fifth night in Columbia, and I must say, there is a constant familiarity in every town. Each town's discoveries are simple in nature, but you need to prepare yourself to look deeper than just appearance. The changes that take place in our country to *modernize* often cause the displacement of simple pleasures. So, walking into each town center's hotel, you really are transported back in time. You will not find a Holiday Inn or Motel 6. At least, I was determined not to see any, and we never did. But the hotels we did reside in for just a night were always charming and presented a simplicity in their rooms of yesteryear.

When you step up to the front desk, there proudly displayed is the hotel's computer system. Being 1995, these computer systems were definitely your green screens. There was always background music playing throughout the entire building. The music systems did become a mystery to us. How well it was set up and so interesting that every town center hotel had one with the same music playing. Yes, the same music.

There was a mighty fine salesman making his way ahead of us on the Pan-Am Highway, stopping along the same route, providing some very nice updates to these hotels. We believe every hotel proprietor received an added bonus—three factory-painted mountain scenes. Must have had a high production run on them. Again, every town center hotel's lobby had the same painted mountain scenes. At one point, Robbie and I would bet $1 as to whether we would see the same paintings each time. I always lost, rolling my eyes as we entered, and Robbie throwing his eyes up with a sudden wink to me. Needless to say, I finally stopped betting after the third time of losing. Who would have ever imagined little town hotels would have such similarities! It was fun to observe this occurrence.

And the hotel proprietors were flattered to know they were *in step* with the other towns since they didn't have the chance to travel. That is also important to realize, people outside of the US of A are not as driven to travel, as we seem to have become.

We find ourselves (happily) in locations that enjoy simple pleasures as a priority and do not displace simplicity with automation. TV is not prevalent, except you might find one in the guest lobby or bar where many are gathered around watching a soccer game. The town square is always alive and full of people walking about and shopping. The restaurants are full, and people are also outside enjoying the fresh trout and vegetables. Great coffee, wine, and all in the spirit and rhythm of the Colombian music that is playing either live in the square or provided through a restaurant's sound system. Just enjoying the day and night. Every town is like this, and it is joyful.

From Cartago to Popayan it is just over two hundred miles.

I must say I was totally blown away by Popayan's appearance. The town is lovely and very historical in nature. I didn't know there was a colonial history in Columbia. Yet, this lovely city of Popayan showed every sign of architecture from a different era and country. So I, of course, drove us into the city, and as we passed a lovely hotel, we said, "Let's keep driving into the town center."

It was amazing and great to see such very different city architecture. It just shouted out its difference and appeal. We agreed to find a parking spot in the town center. Gosh, this was going to introduce more than I could imagine. Plus, it was dinnertime, so what could we find? Robbie was already talking about his cravings for mashed potatoes and peas and sweets for dessert. I didn't really have any food cravings of notice. I parked Velveeta, and of course drew a few looks, as always. This was our first cosmopolitan, yet colonial-looking South American city.

And now, I must admit, I was craving a cheeseburger! Of all things, and I really didn't know it until we had walked a couple of blocks. Robbie's comment, when I told him about my search for cheeseburger was so typically British, "Darling, you are truly an American." The way he said it made me know this was definitely not good breeding!

My comment in reply, "Oh, ta ta."

As we walked through the city, I had a sighting! I was undaunted. I simply looked over at Robbie and said, "There it is!"

We entered the restaurant, stood in line to give our order, and then made our way to a table with a window view. This was the most delicious cheeseburger ever. I had not had a cheeseburger since leaving Portland, Oregon. It was just a plain cheeseburger. No condiments and it was juicy and medium-rare. Lovely. Unfortunately for Robbie, there were no mashed potatoes and peas, but he did have a burger and fries along with a sweet. The rest of the evening was pleasant and the hotel we found was very different from our countryside town center hotels. It was very modern and we could have been anywhere in the United States. There was no need to inspect the room before registering. And the morning brought Robbie lots of hot water for shaving. I, on the other hand, still had cold floors to cover up for warm walkabout.

After Popayan, Highway 25 was straight and clear. We had one more night in Columbia in the city of Pasto, then on to the border crossing at Ipiales (Columbian side), and then Tulcan (the Ecuador side).

All in all, Colombia was delightful. Yes, we did have some sorted adventures in this journey. Yes, Columbia didn't exactly provide us with the best step forward at the start. Just about every day presented an element of surprise. But how else are you going to have an adventure if the journey doesn't present surprises? There were far more sights in this adventure than any guidebook could possibly suggest. Gosh, the beauty from its people and country. I wish the books had been nicer, and yet, I'm happy they were not because I was always happily surprised by the kindness and beauty. And when there was distress, I knew we would recover. Colombia deserves better press and more tourists.

We made our way to the border crossing at Ipiales, then seven miles later, Tulcan for Ecuador. Seven days and slightly less than one thousand miles. I would never have dreamed so many adventure memories could be had in just seven days! Fantastic is the best word to describe our Columbian adventure.

Date	Mileage		Number Miles	Location for the Night (or more)	Notes	Days Travelling in Country	Miles Travelled
				Entry into Columbia			
6/20/1995	151397			Cartagena, Colombia	What a horrible experience getting into, or should I say attempting to leave border crossing. The soldier that seemed to think we were carrying drugs into Colombia, took everything out of Velveeta, including emptyinig my purse. Which he demanded I empty it and I told him he wanted it emtpy he would have to do it himself. I stood outside of Velveeta and refused to move. Robbie was furious with me but I could not care. The soldier then emptyied the pursed then stepped away. We finally paid and left. What a horrible 3 hour experience. Welcome to Columbia. We were going to stay the night in the city but decided we had experienced enough and left.	7 days	982
20-Jun	151515		118	Sincelejo, Colombia	Stayed in Hotel Marsela - very nice		
6/21/1995	151759		244	Santa Rosa de Osos, C	The drive thru the country side was beautiful. Up into the forested mountain. Young boys with hose coming down the mountain washing the trucks coming and going. There is a lake at the top that fills the hose. Water abounds. Staying in lovely hillside village. Arrived following a funeral possession (all walking). Able to get around and get into the center of the village and find a hotel.		
6/22/ - 6/23, 1995	151821		62	Caldas, C	Velveeta needed repairs to emergency brake, $25,000 pecos.		
24-Jun	151964		143	Cartago, C			
6/25/1995	152172		208	Popayan, C	Very stately city.		
6/26/1995	152326		154	Pasto, C			
6/27/1995	152379		53	Ipiales, Columbia	Border Crossing from Columbia to Ecuador		

Chapter 7

Ecuador—In Search of the Equator

Crossing into Ecuador was relatively painless. The *transito* was about $5 to cross the border out of Colombia. They simply stamped the passport and took the money. Done! Seemed odd, but then again considering what we went through at the Columbian entry, we are lucky. The Ecuadorian side for entry was all about money, hand out, hand out, and hand out. I was so sick and feed up I said no to everyone. Robbie's words are a bit stronger than mine on the subject of entry "hand out." Robbie's colorful comments were basically about the "hordes of scruffy urchins" all assuring us that they could get us through immigration cheaper than anyone else. Robbie kept calling them "Rotten little blithers." The immigration fellow wanted $20, and Robbie told the fellow his friend was the Ecuadorian Ambassador in London, and he would make mention of this incident. So the official quieted down and didn't require payment. The only payment we made was to *not* have Velveeta fumigated. The smell is beyond bad, and we knew we didn't carry any bugs. Another Robbie quote on the topic of fumigation, "A sheep's dip in the wilds of Scotland would be healthier and a more pleasant experience than having one's trusty steed fumigated inside and out by a moron dressed in deep-sea oilskins and gum boots, and looks like something from the outer galaxy. The sickening smell from the fumes stays with all of one's personal effects for a good twenty-four hours." It took us an hour between the "little blithers" (so unkind but true), persuasion of immigration official to not require payment and paying for no fumigation. Not too surprising that we never received receipts or stamps for entry on Velveeta. Only our passports were stamped done. We finally drove off with the satisfaction that we don't smell of fumes as we drive south on Highway 35.

Our first night in Ecuador was in the town of Ibarra, just forty-nine miles from the border, found us in a terrible hotel. One of those so-called Bahamian-type weekend resorts. There was no water, no shower. There was hardly any electricity. We argued like hell the following morning, but they would not give any discounts. We created a lot of noise and some personal satisfaction out of it all. I want to quickly write about last night's dinner. Just about the entire city was closed down yesterday. We could not find out the reason. It was Tuesday and not a holiday.

We caught a taxi, but every restaurant was closed. Finally after three unsuccessful attempts, we found one open Chinese restaurant for our first night in Ecuador. Seemed very strange but, hey, why not. Robbie ordered soup; he was coming down with a cold. I splurged with a chicken and veggie combination with rice. It cost us 12,300 pesos, approximately $16. We were sitting next to a family of five who were all sharing one dish of fried rice. After spooning chicken and veggies and some rice on my dish, Robbie offered the remaining food to the father, indicating it was more than we could eat and if they would like some. Yes, the kids dug in as did the father. All gone. The mother did not take any. She also made sure everyone in the family was feed before eating a small remaining amount for herself. We smiled and wished them all well.

From Ibarra, we were off to the village of Otavalo. It was just under sixty miles from Ibarra. Arriving early in the day would give us plenty of time to find our way around the village and locate a good place to stay for a few days. We have read so much about this village over the past year and, of course, pronounced its name all wrong. The correct pronunciation is O-tav-a-lo. We wanted to make Otavalo our base and rest here for three or four days. This will give us time to check Velveeta out, as well as enjoy our search for the Equator. Otavalo is supposed to be very close to the historic equator "line".

Along the way, we did manage to visit some of the haciendas that are in the guidebooks. Since we were out in the countryside, maybe that might be a better choice than staying in Otavalo. But once again, the prices were out of this world. Seventy to seventy-five bucks a night. In this so-called famous hacienda "Cucen", it just seemed so pretentious and in such a state of disrepair. The guidebook, *Planet Series*, proved to not to be worth it. Besides, it turned out the haciendas were lovely structures but not big sparling estates. Yes, you can see the countryside, and a few do have lovely acreage, but it's not a large working estate. It's an inn in the country.

Along the way, we were able to fill up with gasoline. Super was rare here, but you would see the signs for sin polmo, which means unleaded. Better than nothing!

Continuing to travel to Otavalo, we checked out the village of Cotacachi, just fifteen miles south of Ibarra. It seemed to be famous for its cheeses. The travel guidebooks said nothing on this topic, but the shops had plenty to offer. After some tasting, we decided there was a reason why the guidebooks didn't mention the cheeses. They were very, let's just say, ripe! We will pass on these today.

Although we had only been in Ecuador for a day, we found the Ecuadorian customs, or maybe it's just this region, were rather unique. Women were wearing a traditional blue-and-white skirt and loads of glass beads around their neck. We had not been about to find out the significance of the small to the large quantity of beads. The men wear very baggy pants that were only calf-length. Everyone's hair was long in length and in brads. All very clean and very friendly. Cotacachi is definitely a tourist village based on the advertising and the shop owners attempting to entice you inside.

While I wandered around Cotacachi Robbie stayed behind and worked on Velveeta. Not a lot of work was needed for Velveeta, just change of spark plugs to the platinum, German ones. Peru is just down the road, and we know we can find a Toyota dealership for Velveeta's care. Maybe three to five days from here. So we put on a new air filter to tackle the dust in the desert for our future travel down the coast to Peru.

It was a brief drive into Otavalo from the village of Cotacachi. We drove right into Otavalo's town center. We parked Velveeta at the main plaza and walked down to the first hotel we saw, Hotel Indio. We once again were up against the budget. This time, the owner was very reasonable. It just so happened the hotel clerk owned another hotel that might better fit our budget. We couldn't afford the fifty thousand sucre for Hotel Indio, but he had rooms for forty thousand sucre, and he wouldn't charge the 20 percent tax. The hotel owner drove us to the Residencial Indio, just around the block. The hotel has seen better days, but by this time, so had we! The hotel was just off Plaza de Ponchos, where the market is held. The hotel owner took us upstairs to check out the room. Since our terrible experience from last night was still lingering, I found myself immediately checking out the bathroom. The light didn't work, no water in the faucet and no hot water in the shower. The owner simply said not to worry, and that he would repair everything right now. He was off for some tools, and in no time, all was repaired. We joined him downstairs, and we signed in. Robbie, of course, regaled our travel adventure and introduced us. He introduced himself (Alfonso) and then drove us back to Velveeta. Robbie complimented Alfonso on how nice it was to meet someone that would work with us on the price. Alfonso had a broad warm smile and was very pleased and of course, so were we.

Unfortunately, our first night in town was terrible, not due to the hotel but due to the music blaring through the night. We learned the next morning there is a religious festival taking place.

It is Festival Time in Otavalo.

The music went on until five thirty in the morning. That may have been the explanation for last night's closures in the city of Ibarra. The morning brought the sound of buses, and then the cockerels could be heard crowing. It was impossible to sleep. Are we still in Otovalo? Between the roosters and buses, I wasn't sure if we were in the city or on a farm.

Anyone Seen the Equator?

We decided to make our first day of rest a search for the Equator. It was not funny at the time, but we could not find it. We drove around and never saw a sighting. We did discover lots of interesting outcroppings of rocks and the rock strata that was so diverse in color and texture. Warm red and pink colors then large streaks of black. Odd, but I suspect it must all be the after effect of the volcano eruptions of millions of years ago. We were surrounded by volcanoes that are no longer active (at least not while we are visiting). After four hours of searching for the equator and finding nothing but interesting rock formations and caves, we returned to Otavalo. What turn did we miss? I revisited the maps that night so the next day we would have a successful adventure to the equator.

We pulled out the maps and discovered we had not taken the correct road. Duly noted, and tomorrow would be a successful day. Of course, I didn't say anything about the fact Robbie was driving, and maybe I should have been behind the wheel. Way to knotty at this point! Even if he claimed he had flown over this area many times in the past, it's one thing to be in the air, looking down versus trying to find what should be right in front of you. I simply brought out the good ole gin and tonic. We marveled at the most unusual mural painting of a woman on the wall behind the bed. Robbie set back against the headboard and threw up his arms, ginny-winny splashing about and simply yelled out, "Yahoo!" Talk about a funny sight, I have a photo for you on that one!

Friday, we have had our breakfast, and with maps in hand, we are off to find the equator. I state we will not return until our discovery is successful! I felt like Lewis and Clark but in the wrong hemisphere!

This time, we would drive to the village Buena Esperanza. This is the location for what is considered the authentic monument of the Center of the World (Mitad del Mundo). There are other markers north of Quito, established when the GPS was made available and placed the equator in a different location. But then again, the equator does not stand still. It actually fluctuates based on the earth's axis. So the French equator discovery of 1735 could very well have been crossing Buena Esperanza. We decided it did and made it the spot we would celebrate.

This site is fifty-five miles south of Otavalo. Buena Esperanza is very small village. Hard to believe this village would have had hundreds of visitors before changing the location of the equator.

Constance Happy to Find the Equator

We each had our photo taken with Velveeta and the globe of the world as a backdrop, as well as photos with the equator line marker. I knew we both looked the part of adventurers exploring the great unknown. We smiled inside and out and laughed for hours on end. Just walking around the monument made us both giddy. As the day wore on, we knew we needed to return to Otavalo, but we really felt we had met an incredible journey goal. For goodness sakes, we were standing at the historic equator. So we pulled open the back of Velveeta and sat together looking around at the entirety of the markers. We talked about all we had accomplished to arrive at this very special spot.

The months of planning and preparations. The weeks of packing and along the way rebalancing. Driving and driving and never thinking we should stop and turn around. Never, not even after the most frustrating incidents. Reflecting on the various frontier crossings and episodes of pleasure and dismay. All in all, there we were. The two of us and Velveeta. We had driven almost eight thousand miles, and we were all doing pretty well at this point. We even gave Velveeta a big hug to let her know how much we loved her performance. As much as we didn't want to leave our place of glory, we must return to Otavalo and prepare to leave and continue on our quest to reach Argentina. Our feeling of pride and accomplishment was almost overwhelming. We were also filled with knowledge and recognition that we could accomplish anything together.

Constance and the World Robbie in the Distance

Saturday, July 1, we had planned to leave Otavalo and head south. And as luck would have it, or no luck as the case may be, it was market day. As we walked to get Velveeta and drive back to the hotel to pick up our belongings, we were surprised to see Velveeta was blocked in. There was no getting her out. The roads were turned into a market place. Oh well, so we'll stay another day.

I decided an outdoor market would be just the adventure for me and my camera. Robbie was going to relax indoors and just read. I grabbed my trusty Nikon and was out and about.

Constance at Otavalo Market

The colors were lovely throughout the market. Even the clouds were constant companions with the blue sky peeking in throughout the day. You could see the mountain peaks in the short distance. I didn't realize just how far up in the sky we were until I walked to the outdoor market. We were surrounded by volcanic mountain peaks at 8,300 feet. Lovely!

Is She Buying or Selling Those Chickens?

The market was full of people and animals. Women carried hens in a cloth bag slung over their back. They made their way to find vegetables for trade. Difficult to say who was trading what. A lot of woven ponchos and other woolens stacked in piles. Each pile had its owner sitting on the ground and leaning against his property. You can lose yourself in the market and just blend into the crowd. What really stood out was me. I stood at five foot two, I would be considered a tall person. No one, man or woman was, as tall as or taller than me. That's a first for me, wow, a giant. As for my red hair, it loudly spoke that I was not from this country. Nonetheless, regardless of my height and hair, everyone was pleasant. No one was disturbed by my photo curiosity other than a woman with a British accent coming up and admonishing me for taking photos. Taking photos might cause the village people to think I was taking their souls. Well, I stated I had already made the damage to quite a few at this point. I did inquire and was told photographing villagers is not a sin unless they were expecting money. No, none offered and none requested. No problem, have fun and enjoy your photography. I felt like my spirit was lifted, and I was on my rounds to explore the market. And what a market it was. I was really fascinated by the piles and piles of fabric and yarn. The Ecuadorians were definitely farmers. Whether it be meat at the butcher's table, chickens getting their heads chopped off (yep, right next to you walking by) or veggies, clothes, bags, you name it, and they made it or grew it. Lively and pleasant. What a surprise and delightful day.

Here's Your Chance to Sell or Buy!

Our day of departure arrived, and we now must leave Otavalo and be on our way to Santo Domingo, just 150 miles. Unfortunately, we missed the turn off (Highway 30) to go over to the coast, and we finished up going up over the mountains and finally stopped for the night in Riobamba. Well, a bit farther at 180 miles from Otavalo and still on Highway 35. Robbie was absolutely infuriated, but now it really didn't seem to have made much difference. We did get to see a great deal of Ecuador on this route. So much of forests were gone. I give tribute to the farmers (men and women) who work the land. It is a hard life in the mountains. Unfortunately, the tourists in their touring buses, out for a picnic lunch leave their filth on the roadside. We continuously saw the countryside was missing its great forests. Very sad indeed.

We got into a little hotel in Roibamba. The water was completely turned off, and the hot water was virtually nonexistent the following morning. Plus, no food to be found, so off we went and on to the next stop Guayaquil.

We had tremendous mountain driving, and I'm probably exaggerating, but the grades seemed to be between 20 and 30 percent. We were cresting between 14,000 and 15,000 feet. Finally up through the rainforest. Tremendous, absolutely tremendous forests. We couldn't see more than six feet in front, sometimes less. The stupid buses and trucks, they just barrel through with no lights on whatsoever. We saw retired people, senior citizens in a rented buses, sightseeing. They stopped on one of the mountain passes on a little green area there overlooking one of the beautiful valleys. They brought out their box lunches. Seriously, just sitting on the side of the road picnic at hand.

At one point, we did have a bit of difficulty with the route. The road all of a sudden turned to dirt, and we were on the side of the mountain, wondering if this was the end. Fortunately, around the corner, we were fine again. Maybe there had been a washout of the road at some point and not repaired. No signs but no road. But we made it through, and we're barreling down to Guayaquil. I must say, having missed Highway 30 did cause a lot more mileage. We really wanted to get closer to the coast and thought the best plan was to get over to Guayaquil for the night. We made our way off Highway 30 to Highway 70 to take us to Guayaquil. The next morning, we would back track to Highway 25 and head south for the coast.

But first, let's get into Guayaquil for the night. Routine at hand, we exchange seating arrangements, and I take us into Guayaquil. It's early at only three o'clock in the afternoon. I don't really want to say much on this topic, but even Robbie was extremely happy I was behind the wheel. We didn't get into a hotel until after six o'clock that evening. This was the craziest laid-out city I have ever seen in the world. Nothing conformed. Hours and hours going around and around trying to find hotels and following directions from the stupid guidebooks. At one point, we just gave up and said we would take the first one that we see that would look decent. We found one and looked it up in the guidebook. It was classified as moderate. All of the windows were knocked out and dilapidated. Needless to say, we did not stay there. We ended up in a $47 a night hotel, which was classified as moderate in the guidebook. You get this 10 percent plus 10 percent on everything. Just plain expensive. Nothing more to say.

Once checked in, we made it to the hotel's restaurant. This was actually a very nice one. We stood out like vagabonds, and Robbie introduced us to the waiter. Another nice fellow. I must say the dinner was a delight. I had my very first individually baked paella. Absolutely delicious and filling. Just what we both needed. Welcome to Ecuador!

Next morning was July 4 and Independence Day back home. We were in a celebratory mood as we left Guayaquil and planned for the border crossing in Huaquillas. Hopefully we would have the border crossing into Peru the day after.

Once again, we became rather pissed off at trying to get out of Guayaquil. It took at least an hour to get out of the city. I did all of the driving today since Robbie had a full day yesterday. Very easy with level straight roads. But once again, absolutely nothing marked. We were continuously asking questions to those we come upon at the gas stations. I must mention that the diesel fumes along the way were incredible. Really bad, really bad. The pollution you can see was really settling in. And once again, the filth on the side was so very sad. We saw the Dole banana plantations along the route. Incidentally instead of using the word *plantino*, they use the word *bano*. But the banana plantations look not very well off. Must be the weather conditions. I would describe them as well worn!

We did back track to Highway 25 and made our way south. We only made it to as far as Santa Rosa just before nightfall—only 126 miles, not what we planned. At the gas stations, we found out there was a new road that was built that would get us directly to the intended border crossing if we watch for the Highway E50 out of Arenillas. This highway should lead us to Highway 1 and down the coast of Peru. We should see if the plan would work out. We would need to be very vigilant and watch for signs or just a new highway that lead us west.

So here we were in Santa Rosa and needed a hotel, and it's late. Finally, we found an establishment, and on checking the third room, I was satisfied even though I didn't check out the showers. Once we got settled inside, I went into the bathroom. Of hell, there wasn't a shower. Please, it was bad enough to admit and tell Robbie. You could only wash so much from a sink, but at least there was water, and Robbie could have his morning shave. I was simply ready to get on into Peru.

Wednesday, July 5, we left Santa Rosa, and hooray for us, we found our new highway out of Arenillas into Huaquillas.

Huaquillas was a very easy border crossing. Both leaving Ecuador and entering Peru. Very interesting point, we were the only people leaving Ecuador and the only people entering Peru. Yet it took two hours. No money hassles though. Just slow.

In closing Ecuador, we found ourselves setting too many expectations on the sites of Ecuador that just didn't occur. Yet it was an incredible country with its volcanic mountains, valleys, hillside farming and villages. Lovely people and welcoming everywhere. Our favorite spot is undoubtedly Otavalo and the visit of the ancient site of the equator. Both remarkably memorable in the very best way!

				Ecuador			
Date	Mileage		Number Miles	Location for the Night (or more)	Notes	Days Travelling in Country	Miles Travelled
6/27/1995	152386		7	Tulcan, Ecuador			
6/27/1995	152435		49	Ibarra, E			
6/28 - 7/1, 1995	152493		58	Otavalo, E	Lovely City. Decided to stay a few days and enjoy.	8 days	781
7/2/1995	152782		289	Roibamba, E	Very few signs to help with the route. In the mountains all day. Riobamba is without water until late at night. Time to shower!		
7/3/1995	153005		223	Guayaquil, E	Hotel Dorado. Had excellent paella		
7/4/1995	153131		126	Santa Rosa, E	Hotel American		
7/5/1995	153160		29	Huaquillas, E	Last town before entering Peru.		

Ecuador Travel Legend

Chapter 8

Peru—Diverse Land

At last, after taking two hours to pass through Peru entry at Huaquillas for immigration and vehicle inspection, we are now continuing our travels south. I'm not going to bother to ask why it should take two hours when we are the only ones coming into Peru or going out of Ecuador. The proceedings were smooth, just slow. So leave it alone! I also want to comment that at the time, we were not aware of the border conflict between Peru and Ecuador. We learned of it all much later. We just stuck to the highway and made our way going south.

Peru Coast—out having fun

We now have the Pacific Ocean in our view and the fabulous smells of the fresh ocean breeze. How very grand and absolutely refreshing. We have been land-locked since leaving the South American entry port at Cartagena, Columbia, just over two thousand miles ago. Yes, I do believe we deserve a bit of ocean breeze and change of pace (hopefully, that is possible along this interesting route named N1).

We plan to spend our first night in Sullana (just 159 miles southeast from the border entry), then tomorrow, we shall continue our southern route with a brief venture east and then back west to the grand Pacific coastline.

As we enter the reasonably sized Sullana town, we immediately look to find fuel for Velveeta. Unleaded is available, albeit at $2.25.

(Robbie constantly grumbles over the ever-increasing fuel costs, but Velveeta does not. And she is more important!)

We find a nice hotel, which (like many hotels in remote locations) has a secure parking lot for Velveeta for the night. Hotel La Siesta, home for the night, is reasonably priced and has nice accommodations as well as a restaurant. A nice unexpected surprise. Lovely people and a very clean room.

The first night in Peru calls for complete focus, *gauging* ourselves to meet Peru. Yes, meet Peru. You can't just drive through a country without first coming to understand its characteristics. What can we learn in order to minimize the risk of absolute *surprise* every day? Well, we love a good scenic surprise, but not really a road surprise. This is our practice of exchanging the last set of reference books and maps (Ecuador) for the new country's set (Peru) so we can prepare for enjoyable and relaxing discovery and adventure of Peru. We have learned that it's a good practice because looking into the details ahead (highway maps (usually two on hand), guidebooks for towns, which then inform us of for what degree of accommodations and the must-have bank directory), we can select and coordinate back to the map for mileage.

Depending upon the town's accommodations, you surmise what could be there for fuel and food supplies. Our desire is to stay on N1 because we very much want to hug the coast for a while before heading inland. We need to determine if that will be possible considering the priorities. You see, I never carry too much cash. No matter how many reference books and guides you have on hand, you don't know who or what will show up along the road. That was a very significant discovery through Central America and Columbia. I keep just enough for a two to three-night stay.

Again, it depends upon what can be found in the bank directory coordinated against guidebooks and map. We start our night of work not only by exchanging our reference materials but also, we pull out our travel plans created in Oregon. I would say, 90 percent of the time, we are good to stay with the original travel plans. Only the town's accommodations have been the 10 percent not meeting the plan. This, in turn, can mean a negative impact on fuel and food supplies. For Velveeta, we were always 100 percent okay. Only the accommodations and food suffered the 10 percent that missed the mark of expectations first set in our plans. Not too shabby! Remember, in 1994, using the internet to look up the locations and accommodations was not the best reference tool at that time. Smartphones to carry along the way, nope! Our references come strictly from the travel guides and my trusty Citibank reference book.

Looking over the highway map we discover the N1 Highway (which was our choice after talking with several people in Ecuador) is unlike those we are accustomed to across all the countries traveled to date.

N1 not only can take you south, but you have designated points that you can choose to go south or east or back north. All marked N1. So it's not just a highway going north and south, it also goes east and west, at the same time at different points along the way with the same name. The eastward N1 starts leading you toward the Cordillera (backbone of North, Central, South America, and Antarctica) which includes the entire Andes mountain range. Not ready for that yet, we just left the mountains! We give each other that well learned the *roll the eyes, here we go* look, which means, *be cautious and don't get caught in taking an unwanted turn, or at times going straight when you should take a turn.* This will definitely require both of us to be vigilant in watching the highway and taking turns at the wheel. Do not want to become overtired at the wheel or in the passenger/navigator seat.

We also need to make sure we know what towns are next up and noted in the travel journal in case the highway markings get out of hand. That is, towns are not being named on any sign posted along the way, but seeing other names that will be reference points. Yes, that is what happens. I also make sure to notate those towns that are not too distant that have the very necessary ATMs, as well as possible overnight rest stops. When I state *rest stop,* I mean possible inn or hotel, not a *rest stop* like in the states. No, you do not spend the night out on the roadside. That definitely can be a ticket for consequences you will not want to experience. Difficult enough during the day!

I have not mentioned food because you really can't always depend on the guidebooks in these little towns and villages. If there is a restaurant or merchant with food, then we are in luck. And most of the time, you are in luck. You just need to adjust your expectations and realize it may only be bread for the day. And that's okay! And if nothing that night, not to worry, food will be found the next day at some point. We never had to go more than overnight or the following day by late morning without food and water. Just keep your eyes open and inquire. People are always helping us when we ask. Everyone must eat!

The next day (July 6, 1995), we head south and go toward Sechura (approximately forty-five miles away) in order to come upon the Pacific Ocean. The morning is lovely, and the ocean is fresh with that lovely salty air scent. Missed you, Pacific Ocean, since leaving Panama City! There is a tremendous amount of fishing activity taking place. Peru has a lot of road construction underway, so it does make you wonder about their future plans for commerce. The beauty is brief, and we must head inland for a short period until we reach the small town of Monsefu, another fishing village. So much life and so much ocean! Monsefu is just north of Pascomayo. In a couple of hours, along N1, we enter the coastal village of Pascomayo. Home for the night.

The Pascomayo village is right on the Pacific, but there are no fishing boats to be seen. It has a great pier, but no boats. Where they went and why is a village mystery to us travelers. Maybe the fishermen are several hundred miles out in the ocean fishing. It is a clear day, and we can see out at least fifty miles out, but nothing in the distance. Well, the next couple of days should bring more fishing life to us.

I will say, Pascomayo village would attract a person to buy a place right on the beach and enjoy time writing and visiting. The homes are like townhouses, or row houses, with lovely wooden accents of color. I'm sure there must be fishing boats out there somewhere.

Dinner brings us great food. Absolutely yummy empanadas filled with meat, onions, hard-boiled eggs, and great hot spices. The cost is so little, only one sol (about forty cents). One is plenty for dinner and if I ate two for lunch and I would not need another morsel for the rest of the day.

Feeding Velveeta is quite another story when it comes to cost. She is not having a problem with the unleaded octane level, but it is giving us heartburn with the price. We have been experiencing 5.46 and 5.65 nuevo sol per gallon. With the rate of exchange at 2.25, the price is exceeding $2 per gallon. Today, we spent $40 just for fuel. This undoubtedly explains why the taxis are three-wheel buggies. They are really cute. A motorcycle front, but the rear is a carriage that can comfortably seat two. Most are open, but there are a few with a convertible top. These taxis are everywhere in the towns and villages. There are also taxis that are only motorized with the power of the human pedal. The people of Peru will not be stopped because of the high cost of fuel!

Our travels will remain along the coast as we drive to Huarmey (just over two hundred miles). The cost of fuel prohibits our extending our trip inland. It is unfortunate that the sights of Machu Picchu and Lake Titicaca will not be ours. We will remain along the coast and enjoy visits to its many coastal villages. Lots of life in and around coastal villages.

The drive to Huarmey is filled with surprises, and I do mean surprises! Not a bad road, just a bit unusual. One particular segment of the highway suddenly became very wide. Having been trained as a pilot and experienced many take-offs and landings, I must say I was astonished when the runway markings suddenly appeared. So astonished that I stopped just before the runway with bold markers that let you know you have reached the end of the runway and it's time to push the throttle, all the way for full speed. I looked across at Robbie. Robbie has been a pilot for, well, let's just say, over forty years. He set straight up and steadfastly starred down upon the road. He looked across at me, and with a huge grin, rolls down the window and shouts out, "CLEAR, ready for take-off."

I pushed down on the gas pedal and off we went. Faster and faster, as if at any moment, we just might have lift off with wheels up. But, gosh darn it, no. The runway continued for a few miles, then at the end of the runway, disappears as suddenly as it first appeared. I must say at that point, we were both laughing hysterically! Robbie was patting his knee up and down and yelling, "Bloody hell, love it, what, ho!" It was absolutely the most exhilarating and perplexing three minutes of our trip! Why on earth would you need a runway going toward the middle of the desert unless you knew of a need? There wasn't an airport in sight. No buildings nor another road. But it was a hell of a nice runway. Another travel adventure mystery to someday solve.

We drive beyond the runway but then experience more road construction. The workers use wooden blocks to the square of sections of the road, then pour the cement, and finish by hand. Have not seen that method applied in a very long time. Considering the method, the roadway is in very decent condition, only about forty miles were badly torn up, but not being neglected. Workers are all around.

Then suddenly, in the distance, we see a fellow on the side of the road. As we come upon him, he is just standing there with a smart cap on, a white shirt and tie, and his bag sitting by his side, on the side of the highway. His hands were folded neatly in front of him, waiting for a ride. This occurred in the middle an oil field where the old mules were pumping away. A few of the oil mules were not pumping, but the majority were steadily working. To this point, we had travelled for miles and never once saw a bus before or after we saw the poor chap patiently waiting on the side of the highway. Hope someone picked up that poor ole soul. He really looked so dashing!

We end up driving through salt flats, which were most interesting. Talk about a diverse land. Lower desert, salt flats, air runway, and oil fields. Attractive in so many different aspects. Never really thought Peru was this diverse.

Driving into Huarmey, we find a small hotel for the night, Hotel Santa Rosa. We shower and go horizontal for the night. Tomorrow, we will see what we can do to pass around Lima. In many ways, we would enjoy a *fly-over* view of the city but would just as soon drive around to avoid a three or four-hour race through the city as we aim toward Pisco. We are thinking Pisco should be fun since it does have a spirit named in its honor. Pisco Sours. We shall see!

Lucky us, we bypassed Lima. Thought it would take us three or four hours, but no, just twenty-five minutes. Boom! Through it. Nice freeway and not a lot of traffic, considering the size of the city.

We arrive in Pisco for the night and unfortunately tonight (July 8), the town is full of drunks. I really hate to put it that way, but what on earth is going on? The overnight stay will be in a horrible little hotel, which is a disgrace. It is without a toilet seat and filthy, so I must clean everything. And wonders of wonders, what do I discover? No hot water! Grr!

At this point, you are wondering why our introduction to Pisco is so off-putting. And why we didn't follow our rules for selecting a hotel? Well, there were no nice hotels with a vacancy in this city. We went into several hotels all around the downtown. Why no vacancies? Because the town of Pisco is holding a weekend religious revival. Tada!

We should have known something was taking place when we started seeing some very unusual traffic coming from Pisco after we passed through the town of Chincha Alta. Just through Chincha Alta, the highway starts taking a turn toward the ocean and down a relatively small elevation of just over three hundred feet. (The town of Pisco is right on the Pacific Ocean.) As we start driving down, we are suddenly met with a remarkable number of oncoming cars and a very peculiar site on each. Every car had a mattress tied on its top. I thought the next town must have been having a mattress sale. What on earth would cause all these cars to be carrying a mattress?

As we drove into the town of Pisco, we could not believe the number of cars everywhere. What on earth was taking place? We decided to find a place to park, close to the downtown square. Lucky enough, we found one that Velveeta could fit into. Our walkabout helped us to discover the town that had an on slot of people because of a religious revival given by a very famous evangelist. Wonderful! Every hotel we entered had the same message, "Sorry, no vacancy."

Finally, one hotel clerk suggested a place that he was certain would have a vacancy. Upon entering the *recommended hotel*, we immediately knew why it would have several vacancies. There were more drunks outside than we had ever come upon. The smell inside was disgusting, but we ran out of options. So we took the only vacancy available (at least that is what the clerk said). Oh, my goodness. As previously mentioned, I found myself cleaning down the entire bathroom. We slept in our clothes and did not turn the bed full down. Just slept under the top cover. A horrible night with drunken singing until the wee hours of the morning.

The next morning, the town of Pisco held its weekly Sunday morning, military parade. A thousand school kids and about three drunks. Oh, boy. Now, I will admit that on Saturday night, we did have a couple of the town's famous drinks, Pisco sours. Pleasant enough, but not enough to stay drinking all night like those who insisted on singing through the night outside our window!

Leaving Pisco, we are on our route to Nazca, which will not be a long drive at just under 150 miles. The further south we go, the more desolate it becomes. The roads are straight but do not have good camber. Probably no heavy cement under the pavements.

The roads stretch out before you, and you see for miles and miles and miles. At this point, we have driven just over a thousand miles in Peru and have only experienced a few bad spots in the road. As we traveled along the Pan-Am (N1), and at times must travel inland, east. The desert scene stretches far ahead, and you can see the rising of the Cordillera, the backbone of the Americas. The Cordillera is covered in a haze. The heat coming off the desert, meeting the cold of the mountains. The mountain tops simply disappear in the haze. To our west, the desert disappears, trailing down to the coast, giving us a refreshing glimpse of the Pacific Ocean every now and again.

Peru Long Road Ahead into Nothing

Driving through the desert is just plain boring. At times, I find myself getting lost in the distance, staring at the mountain range. Almost exhausting. Sand, sand, and more sand. Flat, flat, just plain flat. One break, we had in the road today probably went up to three thousand feet, no more. Went through a hand-hewn rock tunnel, which made me smile because certainly, the road engineers like working in the desert, but what happened to blasting? This tunnel had been hacked out by pickaxes. And how those trucks got through, I have no idea whatsoever.

Then when we least expect it, something amazing occurs. There are no cars and there are no homes to be seen in this desert wilderness. Suddenly, thousands of chicken ranches and coops. I don't know who eats all of these chickens because there are no towns to be seen. But there must be millions of chickens eaten somewhere in Peru.

On the evening of July 9, we settle into Nazca hotel named Hotel Monte Carlo. Settle is probably too strong of a word to use. Nazca is the home of much history regarding the Inca. There are earth drawings, shapes carved, and lines in the fields. On the way, we passed by a field in which a fellow was waving us down. We stopped, thinking he might need help. Instead, he wanted to know if we wanted to see the hand coming up out of the earth. We said no, but I did get a photo. Not sure if the hand was recently built, or if it has survived for one thousand years. The Nazca culture dates back to 100 BC and flourished for several centuries. I certainly understand how these formations could continue to exist for centuries. There is no rain.

Peru Hand in the Distance

Why Don't We Drive from Portland, Oregon, to Argentina?

As I was saying about the overnight stay in Nazca, we never really settled in for the night. I had to speak with the front desk several times to get the hot water pump working for the shower. The dust was beyond unbelievable. More than I experienced in the Egyptian desert. We found a meal, finally, a hot shower, and a clean bed. We would be good as gold in the morning. We were exhausted from the night before in Pisco, plus the long drive through the desert.

At this point, I would like to point out; we recognize we are not tourists driving about. We are plain and simple travelers. Travelers that must deal with the everyday problems, just as though you were not traveling. The problems are the same, maybe to a lesser degree, as the people who live in the village, countryside, city, or mountaintop that we are passing through.

If the cost of fuel goes up, you eat less. If the produce in the market is bad, you change your diet for the day. You have a different bed every night. The process of selecting the roof over your head requires a checklist that is constantly under modification because there are items that you cannot assume will exist. When any of the items are missing, you just begin to understand how they can do without.

Why should we find after checking into the hotel, that we must buy toilet paper and soap from a vendor outside? Even to get a couple of towels may require a stroll down the street. And don't leave anything behind that you purchased because you will eventually need it again. Actually, many who reside in the town are better off than we are in that they do have constant knowledge of what is available. But then, we know our time, wherever we lay our head, is limited. We wake up and drive on to the next village, making sure we have our list.

This is a slightly embarrassing topic to present, but here I go. No matter how hard we try to find a large enough bush in the desert for me to use for shelter, during those necessary moments of relief, a bus drives by. I'm not wearing a skirt at this point because the weather conditions are too cold. If you quickly pop up, they see even more. If you stay down, I will think I just might have a chance of not being seen. WRONG. We drive for hours and never see another vehicle on the road; I drop my pants behind a bush, hello! The truck drivers honk. I was many yards from Velveeta and the roadside, and I seem to be a beckon in the desert.

At one point, I decided to change my strategy. When coming upon a tight bend in the road with scrub and brushes, this seemed to be a better choice. I selected a spot. Surveyed the spot from outside of the bushes from all angles, even from within. Perfect. Moving quickly, just at that point of no return and what should appear, a tour bus moving slowly *up* the incline of the curve. It moves around to my backside, I am peering through the bush, and yes, the fellow is standing up at the window, pointing out at me. Had not considered the differences between the *high* road and the lower valley. I'm concerned that I'm now appearing in many Peruvian family photo albums. Disgusting. Remember that!

Today (July 10), brings a very long day as we drove just over 350 miles to Arequipa. It wasn't nearly as boring because the coastal drive is so uplifting. Stunning views. Whether it is the out cropping of large rocks in the ocean with lots of bird droppings (guano), that are apparently collected, and the rocky shores where you can see people daring to take their small boats out into the rough sea.

Peru Guano Island

Upon entering the town of Camana, you immediately notice the lovely desert hills in the background. A few twisty roads and hairpin turns, taking us up to probably no more than four thousand feet. Sand dunes all around and a couple of tunnels. All of the tunnels are hand-hewn out of the rock. The cost of fuel has skyrocketed. It was $3.50 per gallon. The fellow pumping fuel had to get a key for the padlock on the fuel tank before he could start pumping. He wanders about for a while as if someone had misplaced the key. Luckily, he found it, and we had a full fuel tank for Velveeta. Velveeta continues to have good fuel and no issues. Three hours after heading out of Camana, and inland back toward the Andes, we finally enter the city of Arequipa. Oh, wow! If only you could see this city. This will be fun, and we plan to stay a few days right in the historic center of Arequipa.

This city will be a refreshing break. We head, as always, to the downtown historic square. We both need a good rest and Velveeta really requires an oil change and attention to gearbox oil. The sand is really hitting her hard. As with us, Velveeta is running rough and needs a good change of oil, bath, and rest.

We found a place to park close to the square and take some time to sit in the plaza. We are entertained by a youth marching band. We were very familiar with the music. Can't help enjoying a marching band!

Arequipa Square

From the square, I found my way to the town's telephone station. It's called the Peru Telefono. I was attempting to reach Citibank in Chicago but could not get through. The operator insisted the transmission was taking place, but nothing on the other end. So as an alternative, I sent a fax. Still nothing. Not a single word. The Peru Telefono folk keep insisting all the transmissions are getting accepted, but no response. Something as simple as reconciling accounts should not require so much time and money. I need to talk to my Citibank personal banker so I can reconcile my bank account and transfer funds into the account and get money available here in Arequipa. But no response. I'll try a different approach tomorrow at a nearby bank that has an association with Citibank. I'm sure that will succeed.

We make our way into a lovely hotel and its restaurant/bar just across from the plaza. Time to just sit, have a bit to eat, a couple of nice gin and tonics. I'm looking forward today to rest for tomorrow. Driving seven to eight hours, a day really does wear you out. I feel like a truck hit me. Robbie said not only did he feel as though he was hit by a truck, but he also felt as though the truck went in reverse and rolled over him again for good measure.

Well, we enjoyed a few Arequipa Gilby gin and tonics, watched the USA get beat by Bolivia, and Argentina beat Chile in the America Cup. The international soccer championship has been a great game to watch during our travels. First started watching in Central America and have watched when we have a TV for viewing. When a TV is on hand, it is in the lobby or bar with many people clustered around. People of every country are enthralled with this game. It is great fun, especially since USA is still playing!

We leave and find ourselves a hotel for a few nights that is just around the corner. We loved the hotel at the plaza (Plaza de Armas) but too expensive for a few nights' lodging. We will return to the plaza hotel for dining and observing the plaza activity. So we settle ourselves into Majestad Hotel and enjoy a shower and clean clothes. We return to Plaza de Armas for dinner and enjoy wonderful dining delight. Funny how we really just enjoy those nice simple meals. The stuffed avocado, not a special. Just the typical item on a Peruvian menu. Lovely. The largest avocado I have seen, with a mixture of chicken. They are not skimpy on the chicken. They must be buying those chickens we saw raised in the desert chicken coops. Lovely tomatoes, hard-boiled eggs, potatoes, onion, and a little something to bind it all together.

With a lovely glass of gin and tonic, becomes the greatest meal on a lovely evening, looking out over the plaza in Arequipa. Cheers in the background from those interested in the football game. This is nice, sitting out on the veranda, overlooking the entire plaza on a beautiful clear evening. Visible in the distance is the volcano mountain called *Misti*. Stands tall about Arequipa at just over nineteen thousand feet. So lovely with its white peak and smaller clusters of mountains aside.

The next day, I'm up and ready to spend the day, enjoying the plaza with my camera while Robbie is off to give Velveeta some tender loving care in the way of an oil change, new air filter, and bath.

In need of papers before going to court

    I find the plaza is full of people today. Most are waiting for their turn in the courthouse. Arequipa has a lovely practice of providing assistance to create papers to present to the court proceedings. A fellow has a typewriter fixed to a short stand. When he sits on the park bench, next to a client, he has a perfect position for typing out papers. He collects his fee and walks about inquiring if anyone is in need of court papers. Better than working inside a hotel all day!

    The plaza is full of birds and people. Lots of activity in and about. I have many photos, so it's time to gather myself and set off for the bank in order to reach Citibank.

    The bank is just off the plaza and I enter, seeking the help of someone who can speak English. Fortunately, this does occur. I explain my situation and provide them with my bank card. They nicely provide me with funds. All of $50. That's it. I indicate that will not do, I'm actually in need of an additional $250 that I can then exchange later into Chilean currency once over the border. They nicely reply that is not possible because Citibank will not allow more funds to be withdrawn. Smiling, I ask for someone to get me a phone line so I can call Citibank in Chicago and reach my personal banker.

Thirty minutes later, I'm connected to Deb and she is trying to keep me calm after explaining that the manager in the credit department decided to freeze my credit card because he thought it was stolen. I said, "Really?"

After all of this time, traveling from Portland Oregon through Mexico, Central America, Columbia, and Ecuador. Now, on our way out of Peru, this man seems to think the card is stolen. Deb quietly responds, yes, he doesn't believe you are driving the distance. My reply is, "Get him on the phone."

After some highly voiced conversation back and forth, my credit card is opened, and I'm able to get the necessary funds. Done!

It's time to return to the hotel, check on the outcome of Velveeta's trip to the *vehicle spa*. We'll freshen up for an afternoon at the plaza, then in the evening, another lovely dinner. We only have one more night after tonight. Just rest and enjoy this lovely plaza and the city sights.

A brief word regarding Velveeta. She is well at 154,593 miles. A complete oil change, air filter, and check on the gearbox and tires. She enjoyed a fresh wash and wax, the front and rear axles oiled, plus steam cleaning of the engine. Beautiful once again, inside and out, and running smooth. Thus her name: Velveeta. A name Robbie wanted her to have because she ran so smooth, just like Velveeta cheese when heated!

As we sit out on the Plaza de Arms veranda enjoying cocktails and the sights at the plaza and the majestic Misti kissing upon the city's edge, we start to reminisce on our travels. We have driven just under 10,000 miles at 9,821 miles. Every country is vivid in our memory. We can now laugh at the frontier crossings; the unexpected commands that have us stop in the middle of the highway, funeral marches, and young boys bringing water down from mountain high lakes and rivers to wash down huge 16 wheelers as well as small 4x4 vehicles. The many historical sites, finding the equator, beauty of the mountains, ocean, valleys, and the many villages and towns and their people, and the funny sight of a car after car with mattresses on top. Never really an ill word or action against us.

Well, albeit threats from border officials aside. I look upon it all as *our time for this adventure through these diverse countries.* Robbie liked to call it as, "May fortune favor the foolish." We have earned a fortune in memories and stories. And some people may someday say we were foolish to have set out on this adventure. We would never say that about ourselves. We saw the good and bad, firsthand. Every country had its own flavor, but everyone wanting the same thing. Peaceful country and the care of its people. It's universal.

We enjoyed one more day and night in Arequipa, then on the morning of July 13, we were off again. We will see if we can have an enjoyable last night in Peru in the town of Tacna. The following day (14), we will cross the frontier into Chile. A new segment to our adventure.

We have been thrilled with the Pan-Am highway and its excellent condition. There are no cars and very few commercial trucks traveling. It is very strange to drive for hundreds of miles and there isn't a single car other than us traveling in either direction. When there is a car you are left wondering how the car was able to hold itself together and make it as far as it did because its condition is close to looking years and miles beyond being driven. We suspect a major contributor to little traffic is the cost of fuel. We must use 95 octanes (we prefer 97, but it isn't always available) and its price has ranged from $5.55 to $6.49 per gallon. It is a killer!

We have not seen any major manufacturing. Only little stalls set up to sell spare parts, a few grocery items, but nothing to really sell or produce to employ people. Don't know if it is the diversity of the country (desert and large mountain range) that is causing the economic difficulty. Desert, desert, and more desert is very much similar to what I experienced in my travels in Egypt. But you come upon the city like Lima and small towns along the way and wonder what has caused the lack of prosperity other than politics, not the land characteristics. Innovation is very much needed to reach out to employ and educate in this very diverse and yet poor nation. The people of Peru are kind and friendly and you know they want the best for their family. Like everyone in every country, we have been able to touch, albeit maybe not intimately, but at least close enough to listen and see them in their natural surroundings.

Our last night, July 13, in Peru is in the town of Tacna, just thirty miles from the border of Peru into Chile. We are in the high desert at about 8,225 feet. It's not a mountain peak, but rather the high desert lying along a desert plateau of the Andes. We have driven just over 225 miles and found a nice hotel. We settled in and decided to treat ourselves to a nice restaurant. And at the hotel front entrance, Robbie hails a taxi. Robbie explained to the cabby that this will be our last night before crossing into Chile and we had been driving from the United States. The cabby's eyes wide and he has a huge smile. Gave Robbie a nod of *got it* and said he knew just the spot. And he did. He dropped us off at the restaurant, El Rancho San Antonio, and promised to return in a couple of hours. Which he did.

Upon entering the restaurant, Robbie let them know we did not have a reservation. The waiter smiled and said, no problem. He escorted us to a lovely table outside in a lovely setting. Flowers and overhead vines with music playing in the background. Just the two of us and another couple seated a few tables away. Fabulous!

When the waiter came for our order, Robbie began talking at length, then Robbie turns to me and simply asks, "Name anything you would like prepared, and it will be prepared special just for you."

Robbie boldly stated he was having his mashed potatoes and peas! I was going to have their fish special and we would enjoy a nice bottle of wine.

What a lovely night. Wonderful meal and wine with music continuing to play softly in the background. After dinner we were up dancing to the music. Simply splendid. It felt as though they had reserved the restaurant just for us to say farewell and you are always welcome to return to Peru.

It had been a very long time since we had enjoyed such splendid pleasure all collected around us. Wow, the lovely time spent in Arequipa and now our last night with this lovely experience. The restaurant staff was lovely and they thoroughly enjoyed our pleasure. If ever I return to Tacna, it would be for dinner at El Rancho San Antonio.

Goodbye Peru, we love your diversity, hello Chile!

Peru								
Date	Mileage		Number Miles	Location for the Night (or more)	Notes		Days Travelling in Country	Miles Travelled
5-Jul	153179		19	Tumbes, Peru	Border Crossing from Ecuador into Peru		9 days	1690
7/5/1995	153328		149	Sullana, Peru	Hotel La Siesta			
7/6/1995	153545		217	Pacasmayo, Peru	Grand Hotel			
7/7/1995	153782		237	Huarmey, P	Hotel Santa Rosa			
7/8/1995	154107		325	Pisco, P	Took a while to find a hotel - Hotel Pisco. They were having a huge revival and cars came in from all over to stay the weekend. Just our luck. I thought they were having a mattress sale. Cars brought in their own mattresses to sleep on in the parks. Of Course, we also bought so Pisco for later.			
7/9/1995	154242		135	Nasca, P	Hotel Monte Carlo (really!) We didn't pay to look at the earth drawings.			
7/10 - 12, 1995	154593		351	Arequipa, P	Lovely city, stayed in Hotel Magestad, but had lunch and dinner in hotel right on the square. Watched men with typewriters help to prepare documents for people going into the court house. Had difficulty getting money from Citibank. The bank would only give me $50 worth. Finally had to call Debbie and have her contact the bank and provide me with $500 worth of local currency			
7/13/1995	154827		234	Tacna, Peru	Just before border crossing into Chile. El Rancho San Antonio. Lovely time.			
7/14/1995	154850		23	Border Crossing	Very smooth exit from Peru and entry into Chile at Arica			

Chapter 9

Chile—over the Top

The border crossing was very straightforward at Arica, Chile. Stamps for exiting Peru, then stamps for entry into Chile. No muss, no fuss. We were on our way in no time and the scenery took my breath away.

I had the good fortune to be first up as a driver for the start of our adventure in Chile. Before going straight to bed, after the lovely dinner in Tacna, we exchanged our reference books and laid out our travel plans for Chile. The first day would be inland, then the highway would return us to the coast. Our route through Chile will be varied, giving us coast and inland views (we hope) of the Andes. I ensure we have ATMs along our route, and we are good to go for a fresh start.

I must say, the start was remarkable. Never before had I seen a land so incredibly vast. The vistas go on forever. Forever because I was incapable of judging its distance. The peaks of the Cordillera were far away to our left, but even with such distance, the mountain's peaks of 21,000 feet look huge. The Rocky Mountains, from the center of Denver, brought in a comparable memory, but even that fell short! No reference point could help me judge and place myself in perspective to the vast layers of the land. The terrain was a gradual soft rolling plain.

The mental picture worked well with my sense of space. Then we reached a position on the road that immediately broke that point of reality and placed me at the edge of a cliff that dropped 16,000 feet. Driving up toward the cliff gave me a visual understanding that the distance between the edge and the other side was a mere fifty feet. I was absolutely wrong. Only upon reaching the apex of the curve, did I see and know that the distance was closer to a mile.

I spoke out to Robbie, "What is happening? Look out around us."

I know my voice was one of disbelief. I didn't want to take my mind and eyes off the road, but I was starting to feel dizzy. I slowly spoke over to Robbie, "Just when I think I am seeing a flat plain I discover it is a side of a mountain. I'm losing my points of reference."

Fortunately, Robbie had many years of experience with the effect of the Altiplano. He warned me of its effect on a number of occasions, and I intently listened to his detailed description of the Cordillera in partnership with the vast Chilean plains.

"No problem," I said again and again. "I can handle the change of scenery."

Now, I realized I was absolutely incapable of knowing how to judge the Altiplano effect. No point of reference against the horizon! I slowed our driving pace and spoke of reference points that would help me figure out where I was in this vast place. Finally, we agreed on one thing. The word awesome must have been born right here in the Altiplano of northern Chile.

First experience with vertigo

So on this cold winter day in July, with the wind cutting through the barren land of the Atacama Desert, I found myself dealing with a fluctuating horizon in my mind and the false impression we had a warm day. Clear blue sky, gorgeous. No, keep remembering you are at sixteen thousand feet in the dead of winter in the Southern Hemisphere. It is blistering cold, and you need to stop driving!

Robbie was trying to talk me through the difficulty I was experiencing. His ever-steady voice and unnerved handling of Velveeta kept us moving onward through this new experience.

I thought my turn at the wheel would help overcome the discomfort. Within thirty minutes, the story changed. "Robbie, I'm losing the road. I can't seem to find my steering point on the road," and I felt panic. My driving had turned into something close to a drunk. I was swerving around corners and totally unable to keep the Velveeta in a proper lane. With a steady and compassionate voice, Robbie repeated his words again and again and stirred me to a stopping point on the road. He slowly pulled me out of Velveeta, placed his arms around me, and held me close so I would not fall. Robbie gave me a steady focal point and pronounced my condition as vertigo. It was a first. I could fly and never experience it, but I was grounded driving at sixteen thousand feet!

Robbie experienced far more in his five decades of flying. He knew this area, from the air. This was his first at driving on the ground! At just over six feet and still carrying his lanky youth, Robbie towered over all we met. His gracious and warm friendly manner relaxed and extended friendships. Even as a stranger to him, he would soon know you and speak directly to the point. From his point of view, you would be better off for it! To him, people were generally too soft and not willing to take risks or hardship. When in fact that was the basis of anything worth having. So it was that he handled even this slight challenge of my wanting to drive off the side of the road in order to pass out or throw up. It didn't make any difference to me, which came first or not at all! I knew I was dangerous at the wheel, I just didn't know how to stop.

Robbie set me in the passenger seat. I had to find the horizon and fix a position in order to stop the dizziness. I had hoped sighting the coast would put an end to this visual torture. The map indicated we were drawing closer. The compass confirmed we were driving west. The ocean had to be there somewhere. Yet, I still could not sense that it was within range. Everything seemed to be on the same plain. Robbie kept saying we were traveling down, I could not sense down. Finally, it came to me with a swift and steep descent. The Pan-Am Highway was taking us to the coastal floor in less than one hundred miles. Suddenly, the awesome Altiplano terrain was replaced by the unending blue Pacific. To me, it was a dramatic and shaking effect because my senses returned. I was coming in for a landing.

The Pacific had nothing visible on her waters. Just an ocean whose peaceful calm appearance was truly anything but calm. We would drive to Iquique for an overnight stay. A short day's drive, albeit, but I was worn out! Every day's drive was clearly planned both in terms of miles, need for fuel, an overnight stay, and timing in finding a bank in order to obtain cash. This plan worked day after day for over ten thousand miles. If there was a change we would tune in and figure something out. We always did. Beyond everything else, we were a team.

Making it to the coast seemed like a miracle. The loss of sense of perspective in judging my *place on earth* was becoming more acute. My need to be at sea level was becoming a feeling of desperation. And the closer its arrival the more it was needed. Just stop and let me out so I can stop moving and focus on a single object and place my body on a given single plain! From sixteen thousand feet to sea level, hurry up! Finally, the coast was there, and I could plant myself on the ground and not fall over. Iquique would be a welcome place to stop and go horizontal!

The morning of July 16 arrived, and I was having my birthday! Onward to Antofagasta! We took a bit more time, but I just couldn't stand the thought of driving back inland so we stayed on the coast.

Welcome committee at Antofagasta's harbor

Upon arrival, I fell in love with Antofagasta's harbor and crazy pelicans! The harbor had a colorful array of fishing boats. A sight I hadn't seen since visiting the fishing villages of the Greek Islands. The fishing boats were painted in such lively colors. The pelicans were in abundance, must be good fishing!

Fishing boats in the harbor

We found a hotel close by the harbor called San Antonio. Once settled, we made our way back to the harbor. The eating didn't turn out to be as good as the pelican menagerie! After eating a dreadful meal, we were wandering back to Velveeta, and Robbie was attacked by a man-eating Pelican! It went right for his leg. Unfortunately, I started laughing and Robbie is trying to figure out what in the world is pulling on his pants. Of course, he looks back at me while I'm hysterically laughing, and he is certain I am the cause of his discomfort.

Oh, no, if you could only see this pelican flapping his huge wings and that giant beak picking at the back of Robbie's pants leg. As if to say, "Hey, hey, I want a hand out. Everyone who comes out that door gives me food, and you may be new to this area, but you feed me one way or another!"

Robbie's pace was picking up, as was his leg. That was one undaunted Pelican! Finally, Robbie was saved by three dogs that went after the Pelican. Of course, it was all happening behind Robbie's leg. Come to think of it, I hadn't seen him dance that jig with such a high step before Antofagasta! Obviously, the Chilean air! Back at the hotel, we both fell ill with tummies very upset. The fish dishes (actually octopus) didn't agree with either of us. Very uncomfortable night for both of us.

When the next day came, we were up and moving. Slow, but moving. The excitement of the Andes was really starting to make itself noticeable. Of course, mother nature was going to help make this another memorable episode. As we drove closer to the point of turning to head east into the mountains, the rain started, and it grew very cold. Finally, we knew we needed to stop and let the weather move out of the region. We had been traveling through the backcountry and loving every small town. It all felt just like a Midwestern area in the rainy winter. We spent the next night in Chanaral and didn't waste any time in planning our departure for the next day and getting into La Serena. It would be another long drive of just over three hundred miles. The drive was stunning with the Andes so clear in the distance.

The town of La Serena is lovely and has a very European feel to it. Great pride and care are seen everywhere. The next day, we were off again, but we didn't drive long on July 19 because of the rains. We stayed at Hotel American in Ovalle for two nights in order to let mother nature calm down, plus we could take care of the laundry. The hotel owner was a lovely lady and helped us in every way. So pleasant and kind.

Finally, just under two hundred miles and we are in La Calera. It is cold and wetter than wet. Driving through the small town, we drove in front of a two-story white hotel named the Grand Rex. It had lovely white-laced curtains in the windows and just spoke to us with a smile. So, in we go. Within minutes of Robbie explaining our travels to the gentleman, we are welcomed as though we were long-lost family members.

Robbie and the owner, Mariano, met and it was as if two long-lost brothers were reunited. Mariano, his two sisters along with his fascinating artist friend, Victor, who spoke wonderful English, had us join them for dinner on our first night. It was lovely. It was a wonderful family party. The Chilean wine and international conversation flowed until the wee hours of the morning. Our luck was truly with us because that night the temperature dropped to zero degrees centigrade, there was flooding in the valley, and the mountain pass was closed because of the extraordinary snowfall! This called for staying under the covers!

For four nights and five days, we actually had a home, the same room, a bed with fresh towels, and lovely meals. The temperatures were freezing, and the rain was still lashing down, but we stayed snuggled, resting our bones. We were long overdue for some very long siestas. Mariano had a great kitchen. Even his empanadas were exquisite. Truly the gift in making them so delicious, flaking, and full of meat and vegetables. And the gravy was added to the lard used for the crust. Absolutely divine. A secret recipe that Mariano won't share, but of course you are welcome to eat as many empanadas you want!

We had the good fortune to visit Victor's home and see his beautiful paintings. Wisteria-filled paintings that brought the scent of spring to the air. Victor had won acclaim in Chile for his artistic skills, but work was in short supply, and paintings were not selling. If only I could, but we had a long way to travel, and it was doubtful that a fragile canvas will stand up to the terrain. So as much as I hated to say no, we did extend an invitation to another dinner at Mariano's so we could once again enjoy an evening to talk and give us all a chance to appreciate the art that came from both Victor's hand and that from Mariano's hand in the kitchen.

All good things must come to an end, at least for a little while. We were itching to move across the Andes, and it appeared that the rains were slowing, so just maybe, we could find a break in the weather. Our goodbye to Mariano was tearful. Our days had been so beautiful, but it was time to part. We said goodbye, but never as though we would never see them again. We had come to truly believe that one day; we would be able to reunite with all of the lovely warm families we had been blessed with touching along our way. The memories would always be very special to us all.

As our travel that day had us turning into the Andes, the weather conditions changed. It was a beautiful drive. The vineyards, olive groves, and surrounding mountains were absolutely fantastic but cold. We stayed in Los Andes, at the base of the Andes where we would make our way up over the top. The hotel we needed to stop for the night was expensive, but convenient to our highway over the top. Roads were currently closed, but we would wait for an opening. Unfortunately, our night of planning was cut short because of a loss in electricity. And I'll tell you what, it is cold in the Andes even under a roof!

Driving from Los Andes, Chile, to the top of the Andes

Driving over the top of the Andes is beyond amazing! What a drive it was! Utterly spectacular! Possibly one of the finest experiences in anyone's life. One not to be forgotten. We went up through the clouds, twisty, curvy road, ice and snow, and tremendous hairpin turns. Oh, it was gorgeous! The ski lift was taking the skiers across the road, and then they disappeared into the clouds. I think we were lucky by not having to pass too many trucks, but the truck drivers were all very helpful. There was one little fiat that looked at least twenty years of age. The most beat-up wreck you have ever seen. Full, full of people, and just grinding his way up, and belching diesel smoke in first gear. Passed us later on when we were on the other side of the border. It was amazing! But I cannot express how fantastic a trip it was. Just fantastic! Twenty-five hairpin turns to the top! You really do feel it is at the top of the world.

Feels like the top of the world

We were there, having met our goal destination of Argentina, 11,512 miles after leaving Portland Oregon, on day 92 of our travel adventure. What a remarkable feeling and set of goosebumps we both had. We gave ourselves a big hug with a good pat for Velveeta, then climbed back inside, and with Robbie behind the wheel, he yelled out, "Tally Ho!" Off we drove out of the tunnel and into the light of Argentina. Bursting with the pride of our *threesomes'* accomplishment! Looking down at the splendid and colorful beauty of the sun-filled Argentina. This travel adventure is just really starting!

Why Don't We Drive from Portland, Oregon, to Argentina?

Date	Mileage		Number Miles	Location for the Night (or more)	Notes	Days Travelling in Country	Miles Travelled
				Chile			
7/14/1995	155042		192	Iquique, Chile			
7/15/1995	155272		230	Tocopilla, Chile	Coastal town. Not very friendly.		
7/16/1995	155395		123	Antofagasta, C	Happy Birthday to me! Went to the marina and greeted by huge Pelicans. I have a great photo.		
7/17/1995	155643		248	Chanaral, C	Hotel Mimi, what a dump. The whole town is depressing.		
7/18/1995	155957		314	La Serena, C	Hostal Crista. Town is lovely and feels very European.		
7/19 - 20, 1995	156013		56	Ovalle, C	Hotel American - lovely owner. She was so pleasant and kind.	12 days	1434
7/21 - 24, 1995	156196		183	La Calera, C	Hotel Rex. Love the owners and staff. Met so many nice people. Had a huge dinner with everyone in the hotel. They had a very nice resturant. We had great empandas.		
25-Jul	156242		46	Los Andes, Chile	Just before border crossing the Andres. We are in the midst of a blizzard. The road has been closed for crossing.		
26-Jul	156284		42		Chile into Argentina		

Chapter 10

Argentina—We Did It!

Argentina Welcome

Over the top and over the border into Argentina, we went through the long tunnel that takes four or five minutes. It was well lit. Both borders were a piece of cake. No problem. All the rubbish about crossing the border with the vehicle into Argentina and what the Argentine Embassy had told us in Chicago was non-existent. We signed a couple of pieces of paper, and they said, "On your way, and off you go. Bienvenidos, Argentina!" This was July 26, three months since leaving Portland, Oregon. Wow, definitely over the top! We did it!

Robbie proudly with Velveeta in Argentina

The beauty of Argentina was seen in the minerals deposited in the mountainsides. Beautiful pinks and purples. Designs in the hillsides that are beyond imagination. Like a beautiful lady's makeup. Lovely.

Argentina's beauty in color

The temperature was now below zero centigrade. Our first Argentine city for the night would be Mendoza. Just 179 miles from the border. Lovely countryside with vineyards galore. We found a lovely hotel in the downtown area. The hotel clerk gave us directions to a very fine restaurant within walking distance. Good thing we walked because after the wine and gin and tonics, well, let's just say we needed a walk home in the cold! Lovely dinner with *superb wine*. The gin and tonics could have knocked a horse over. They were served in very tall glasses, and the amount of gin was incredible! Needless to say, we slept like babies—over the top.

The next night in Mendoza, we took our time to discover the wines of Mendoza and realized how lovely the reds were even compared to a French Bordeaux. Not bitter, nor tannic, just a lovely red bouquet with the meal.

The next day, we traveled through Mendoza's flat plains filled with vineyards. Row after row, reminding me of the country hillside vineyards in France. Much later, we learned the Mendoza vineyards were world-renowned for their lovely wines. But at the time, we thought it was our secret. We were still all smiles from our enjoyment of this lovely town.

On July 27, we were off to Villa Mercedes for an overnight stay in Hotel Dunn. Villa Mercedes is in the interior of Argentina, well, over four hundred miles to Buenos Aires. To my surprise, Robbie suggests that we should not travel to Buenos Aires, but rather head north to Cordoba, the second-largest city in Argentina with just over one million population versus Buenos Aires with over eleven million. Much easier to get about, and again, a beautiful countryside he had seen from the air when flying throughout Bolivia and reach down into Argentina.

Robbie felt the Cordoba countryside would prove to be worth the extension of our adventure. After exploring Cordoba, we could then head over to Santa Fe and follow the Parana River to Buenos Aires. The second-largest river (just over three thousand miles) after the Amazon River. Well, after more conversation at dinner, we return to our room, take out the map, and decide to go for it. Why not. We are here and it's just over two hundred miles away rather than over four hundred miles to Buenos Aires. Let's take a turn north and see the province of Cordoba.

The morning of July 28, we settled ourselves to drive north to Cordoba and discover the city. Just outside the city, we trade places so I can drive us into the city and find a hotel downtown. We first stop at the ATM I've located in order to be certain we have plenty of cash. It isn't until later we discover we are not permitted to exchange dollars and find I need to figure out how to come by pesos later.

The boulevard into the city is excellent and I pull us up to a stoplight. The vehicle next to us honks its horn and the fellow motions to Robbie to roll down his window, which he does. After a few minutes of exchange, Robbie is laughing and the fellow points ahead, and Robbie says, "Follow him."

It turns out; the first thing the fellow says is that he really admires our 4x4, especially the winch. He wants to know if we have just been out on a weekend outing. Robbie explained we drove from Portland, Oregon, down. Robbie said the fellow's mouth just about dropped to the floor of his 4x4. Then he said to follow him and he would help us to find a hotel and get settled. He wanted us to join him and his family for dinner and tell them all about the adventure. It would be the start of a wonderful relationship with the Porta family.

We made our way to Hotel Aruba and the fellow said he would come by at 6:00 to pick us up for dinner. That would give us six hours to settle in for the night. The hotel cost was only $45 a night, but we knew we would need to find something slightly less expensive in order to stay for a while before proceeding on to Buenos Aires.

That evening, we had a splendid dinner with Gustavo's family. We had the wonderful opportunity to get to meet his wife (Claudia), the little boy (Guiliano), daughter (Julieta), Mother (Luisa), and Father (Juan). The meal was huge and splendid. We could not have asked for anything more. Robbie spent the evening talking, while I sat and listened. At this point, my Spanish is good enough to understand what is being said, but not good enough to keep up or speak with the pace of the conversation. We did explain we wanted to continue to stay for a while but would need to find a less expensive hotel and possibly one that was closer to the downtown area. That was always our preference in general locations.

The next day, Gustavo drove us over to the Gran Rex Hotel. Much better price per night plus restaurant just off the side of the hotel. Plus a private parking area for Velveeta to be safe. Perfect. We move most of our belongings inside so we can start to figure out where to take the laundry, repack, and possibly get rid of things at this point.

Gustavo explains, he will stop by later and bring us up to date with some ideas for traveling adventures. Great!

As it turns out, Gustavo was head of a 4x4-weekend adventure club, and we were invited to join an upcoming event that will have a sleepover at a camp in the mountains. Before that adventure, Gustavo wanted us to meet a friend of his who has an evening travel show on TV. Gustavo brings the young man (Carlos Servali) to meet us at the hotel and we all sit down and let Robbie explain our travel adventure. Carlos is excited to offer us an opportunity to be interviewed and shown on his adventure TV show. Plus, we'll receive a copy of the taped interview for a keepsake. We select a day and time for Carlos to come by the hotel and we would then follow him to the filming location. My goodness, we were about to become famous. At least that is what it felt like.

A week later, we have our day with Carlos. He takes us to an area known as Villa Carlos Paz. Breathtaking scenery.

Carlos prepares for filming us.

Carlos has us pull over and he gives Robbie directions to drive so they can come alongside and film us. Then they will take us to a stop that is somewhat mountainous terrain for the interview and filming. My goodness, it was so exciting. Robbie did most of the talking and Carlos had me provide some answers in English. Mostly about my impressions. When we finished, he said the show would be put together this week for the presentation next week. All we needed to do was to find a TV for viewing. Of course, that was offered by Gustavo who had to have us with his family to view. What fun it was.

That weekend, Gustavo wants us all to travel up to a lake and see a very unusual damn. He was not joking. The damn is called San Roque and it has a cement funnel built into the river just in front of the damn. Water coming over and out of the dam in one of the *spouts* is aimed right into the funnel. The water is quickly removed and does not build up. Fascinating construction—very clever engineering. Just trying to get a photograph was a treat.

The evening arrives for the viewing of our interview appearing on national television. Carlos does the introduction and then it all starts with fabulous music as they are filming us driving. Fabulous. Then the interview stops and there we are. It doesn't really take long but there we were. Carlos then lets the viewers know we will be joining the upcoming 4x4 weekend that Gustavo will be leading. All interested viewers should give Gustavo a call so he can be sure that plenty of arrangements are made for the overnight stop. Later in the week, we received our very own copy of the interview, including the exciting view of us driving. Fun!

Argentina Northern Countryside

Upon meeting Carlos, we are introduced to his friends. Thus giving us more contacts in Cordoba. It also gives me a chance to get out and about with others while Robbie is off taking care of Velveeta and meeting with Gustavo about the weekend outing. He also gives me contacts to help me figure out how to exchange my dollars to pesos without causing anyone harm.

Robbie and I spent our days wandering around Cordoba when we are not on outings with Gustavo or dining with his family. We get a great opportunity to travel the countryside up into the mountains. The mineral deposits provide such beauty to the landscape. The mountains give great treats to lovely villages in the valley. Endlessly stunning.

Finally, the 4x4-weekend event arrives. Robbie made sure Velveeta is ready by taking her to a nearby auto shop. No issues, which is no surprise. At over 156,000 miles, Velveeta keeps on smiling.

We pack up some warm clothes for what will be chilly nights in the mountains.

Gustavo explaining the upcoming 4x4 adventure

There are at least ten 4x4 vehicles joining the fun. And I do mean fun. Gustavo has planned the route, and at times, I'm not sure if he really meant for everyone to make it to the final destination.

Four by fours all on top

We had to be pried out of a ditch, which caused us a bit of fear for Velveeta's axle, but luckily, no harm. The only harm was to our worrying nature.

After a full day of driving off-road through the mountain range, we arrived at a lodging, for an evening of food, music, and plain good fun and laughs. Lots of dancing and swirling about. Robbie and I took a turn on the floor to supply more laughs to the crowd.

Time to dance

Then the time came to introduce us to our outdoor rooms for the night. Each family had their own tent under the stars. It was the night sky that brought the most amazing vision of the day. There were no city lights to detract from the stars. The night sky was filled with the most specular view. I had no idea the sky could be filled with so many stars. Gustavo kept trying to point out the Southern Cross, but I simply could not trace the stars. There were so many that it was remarkable. You really didn't need a flashlight and there wasn't a moon either in the sky. Just the glow of millions of stars that were thousands of light years away, simply shining down upon us. Even in the cold, I was warm down to my toes.

Robbie and I settled into our tent. We each had a sleeping bag to snuggle inside and capture our own body warmth. We whispered about the day's adventure, the evening of warm food, new friends, and fell off to sleep.

Morning arrived along with a warm breakfast followed by packing and getting back out on the mountain pass, this time, going down. Gustavo promised we would enjoy an easier route today since many would be dropping off as we drove back to Cordoba. It was all so exhilarating and refreshing. Now, we return to city life and leave the stars above, but probably never see like that again for a very long time.

Back in Cordoba, Robbie and I sit down and start discussing our next plans. We need to look into whether we can ship Velveeta back to the States since we know we can't legally sell her in Buenos Aires. Robbie will start investigating. I'm going to look into options for the return to the States.

After several days of discussions, Robbie brings forth his idea of the next steps. He wants very much to remain in South America. He has spoken to Gustavo and he will escort us to the Argentina and Bolivia border at which point, we would continue to drive to Santa Cruz and explore Bolivia. This was Robbie's home for many years, and he wants to return. I explain my yearning at this point is to return to the States. I'm ready to start up new consulting opportunities that are throughout the United States and maybe in Europe. I have a strong professional background and I'm renewed and ready to advance into areas I've not touched before. I indicated Robbie was trying to go back to what he had, and I want to go forward to new opportunities.

After much discussion over the days, Robbie still wanted to stay, and I still wanted to go, I said, "So let's make a deal. I'll go back to the States and you can stay and return to Bolivia. I'll annul our marriage. Then Robbie, you are free to return to the life you want. We married because we wanted to make sure if problems occurred at the border either the American consultant or the United Kingdom would come to our aid. But, as it turns out, that was never needed. We are good friends, and we want the best for each other, so it's okay to say goodbye and be on our separate ways."

It seemed like the most logical next step to me. Unfortunately, the idea still wasn't pleasing Robbie. At that point, I realized Robbie actually had always wanted to return to South America and remain. He never planned to go to Buenos Aires to end the adventure. He wanted to remain in South America and return to Santa Cruz, Bolivia. He could have Velveeta and make his way back and live the life he desired. He now knew that was his choice. Not mine.

In the end, we sold Velveeta to Gustavo for $1 and left most of its contents with the Porta family. Gustavo's father was most pleased with the porta-potty. The smile on his face was like he just received the best Christmas present ever. Seriously, so happy. He thought it was simply excellent, especially since it had never been used throughout the journey. I will admit for a split second, I did think if I had used the porta-potty, its looks would have been different if I was sitting on it. Yes, looks would have been definitely different, but not in my favor!

Our departure from Cordoba was a sad one. The entire Porta family escorted us to the airport and sat with us at the gate. Luisa (Gustavo's mother) gave me her rosary and with a hug said it was to bring us safely back to the States and into our future. Gustavo gave both of us big hugs and we all cried with happiness and wondered if we might get to see each other again. I knew I would never forget their generosity, acceptance, and love. Into their home, we were, I really felt, a part of their family. So far from home. That feeling never faltered.

Robbie and I arrived in Buenos Aires and caught a taxi from the airport to our hotel where we would lodge for three nights before catching a flight back to the States. We were both under the weather but really wanted to get out and walk the city. Actually, the walk to explore was me. Robbie was very sick, but I thought the walkabout would help him. Besides, I was not going to leave without seeing the tango performed.

We made it on our second night to find the music and watch the lovely tango to our delight. The city is such a bustle. Music in the streets, shops, and restaurants everywhere. Fabulous metropolitan city. I never mentioned my thoughts of someday returning for a holiday knowing Robbie was still not really happy about his decision to forego staying in South America. I wanted the future. He wanted his past.

We would return to the States and see what the future would bring us. I knew the future could only be new and exciting. How could it not after a remarkable adventure that brought us alive and embedded life inside us through out over 12,300 miles of the Americas. I only had smiles inside and tears on the outside with all of the memories endlessly surrounding me. Our time for this adventure really was the right time in the right places for us. Robbie had once remarked after a day full of fun and adventure, "May fortune favor the foolish." We two foolish soulmates had a fortune in memories.

Nothing more could be said other than thank you to the Americas, and the many people we met, and every place in-between. All equally important and deserving of our undying thank you!

Why Don't We Drive from Portland, Oregon, to Argentina?

Date	Mileage		Number Miles	Location for the Night (or more)	Notes	Days Travelling in Country	Miles Travelled
				ARGENTINA			
26-Jul	156284			Over the Top	Incredible crossing. Blizzard conditions but up we went. At the top there is a tunnel which allows all vehicles to stop for inspection and paper work. Worked out just fine. Hassle free.		
7/26/1995	156463		179	Mendoza, Argentina	The drive to Mendoza very stunning. The colors in the mountain side was so beautiful. Cold but no snow. Must keep in mind ... it is winter down under! At this point we have driven over 12,000 miles. Not bad Velveeta!		
7/27/1995	156638		175	Villa Mercedes, Argentina	Hotel Dunn		
7/28/1995	156854		216	Cordoba, Argentina	Meet Gustavo Porta at the traffic light. He is telling us (I'm driving and Robbie is in the passenger sit) how much he likes are 4x4. Robbie smiles and explains we have just driven down from Portland Oregon in USA. Gustavo's mouth drops and says to fellow him. We do and he directs us to a hotel down town and we become fast friends for more adventures. We are going to stop and stay here for a while.	26 days	848
7/29 - 8/16, 1995				Stayed in Cordova	Went on several trips with Gustavo and his 4x4 club. We were also tapped for a TV show that was aired giving us recognition for our travels. Everyone was always thrilled to hear about our adventures. We loved telling stories.		
8/17/1995	157132		278	Cordova	Decide to return to the states, sell Velveeta (for $1) to Gustavo. The entire Porta family drive us to the airport. Mrs. Porta (Gustavo's mom) gives me her rosarory and wishes us God's speed and safety. We fly to Bueanos Aires. Become very sick and catch a flight to Mimai in a few days. Velveeta had taken us on an extraordinary journey of a life time - 12,360 miles. We said our tearful goodbyes.		
8/17 - 8/20, 1995	Without Velveeta			Beaunos Aires, Argentina			

139

Route thru South America

Epilogue

Now, twenty-five years later, I find the future is still revealing new and exciting adventures. Upon our return to the States in August and a month later, we settled briefly in Florida, and I jumped back into consulting. Set up office and pursued new contacts and business. Robbie continued to think of his past adventures and focused on returning to be creative in the kitchen.

In 1998, we made our return to Portland, Oregon, via a flight to Chicago to say thank you to my personal banker and then continue on to Portland by train. The snowstorms were so bad that at one point, the train had to stop in Washington, and we rode the bus to Portland. Something about when we travel the unexpected should always be expected!

In December that same year, Robbie passed away from the toll of cancer. He had no regrets of returning to the States and lived his last days joyous and treating himself to his usual gin and tonic, followed by a scoop of ice cream just before bed. That was his routine even on the night of his passing on December 9. That night, he also asked me to release his ashes at the top of Mount Snowdon in Wales. This was a place he loved to look out upon the valley. He remarked how he often would turn his airplane toward Wales and fly over Mt. Snowdon to enjoy its beauty over the years. It had been a long time, and he wanted to return.

The following summer, I traveled to Wales and made contact with the Mount Snowdon Railway to reserve a ticket and explained my plans. They let me know they would hold the passengers on the train and give me some time to release Robbie's ashes in private.

When the day arrived, I had a seat in the first car and was escorted by one of the crew. They let me out, and everyone remained on the train.

I privately walked to the top. I looked up and out across the valley and quietly said to his ashes, "May your spirit fly." I then threw Robbie's ashes in the air. Within seconds, a white bird (seagull in appearance) flew out of the ashes and up and across the valley. At that moment, I knew Robbie had found his future soaring in flight.

Robbie soars into his future

I returned to Portland, and I continued my work and a few years later, remarried. Finally, after twenty years of my husband trying to convince me to write about this adventure, I said, "Okay," then added, "Hopefully, it won't be too boring."

I, on the other hand, absolutely loved reliving every segment of the adventure written here for you.

1995, Back in the States celebrating memories

The End

Appendix

Checklist for a room with a roof for the night:

1. Bed, preferably one, not two, or four. (If there's more than one bed in the room, do not touch the other bed(s), or you will be charged.)
2. Private bathroom with shower, sink, and toilet.
3. The toilet should have a seat, but if it is without, then mentally reduce how much you'll pay for the room, no matter how clean.
4. The shower has a showerhead, not just a spout. This is a flexible option.
5. Mirror over the sink for Robbie's morning shave.
6. Hot water in the shower and sink. Must check both. If in a warm climate, hot water is not necessary. If mountain streams fill the water tanks, you better take a shower in the afternoon. And Robbie, shave in the afternoon, don't wait for the morning's cold water.
7. The room is not overlooking a high traffic street.
8. The room does not have high-intensity light out the windows.
9. Soap is not recycled from guest to guest.
10. Clean sheets. Very unusual for this to be a problem. Always check.
11. Ask whether the water is turned off at any time during the night.
12. A fan or air when the heat is high, you must sleep well and feel rested.
13. Bed length is approximate for a person over six feet.

14. Toilet does not cause you to face a wall when seated. It does the clearance for everyone's knees must be met. You can learn to sit sideways.
15. Includes towels.

Acknowledgement

My dear friend of over 50 years, Marla Brumley Ward, agreed to read the first manuscript copy of this book. This was without photos and plain 8.5 x 9.5 layout. After two weeks, I called and inquired… "Well what do you think? Marla said with joy, "I love it! But you need maps of the route. I spent most of my time with the world encyclopedia at my side as I read. I had to know where you were all of the time and I didn't know any of those places. It took a long time to read just so I could try to figure it out. Still can't very well, but so enjoyed the journey."

Marla, thank you. Route maps added and this is for you. You won't have to clip out all of the route maps and tape them together to finally say, "So that's where you were!"

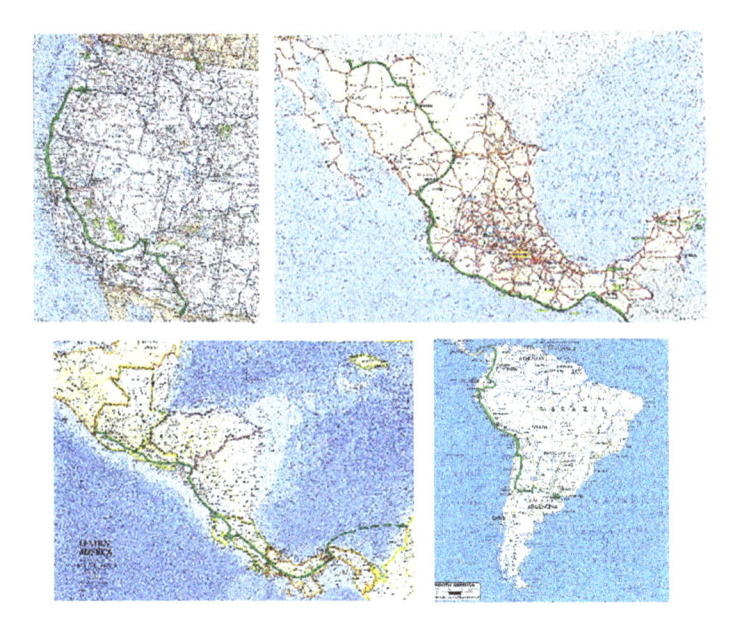

About the Author

This is Constance Glidden Josef's first book but not her first or last adventure story. This story is a favorite and one that she hid away for twenty-five years, only because she felt it might be too personal. Today, she thinks it is a fun adventure story that should encourage everyone to set out on an adventure. Whether it is for a weekend, week, month, or months.

Close to home, nature, hiking, biking, or driving, get out and meet the world and all that is there. People, places, creatures large and small, its foods, and even changes in the stars and sky. Every aspect of her surroundings is different. Her thirty-year career in systems implementation and business operations influenced her to not just explore new ideas and changes indoors but also explore beyond the walls of corporate life. Also as a child growing up in a small village in southern Indiana, the Sunday or weekend outings led by her father really, which she enjoyed, made the first and lasting impression of making sure exploring and adventure stayed in her life.

Join Constance Glidden Josef on a 4x4 drive that departed in April 1995 for four months. Experience pure adventure for over twelve thousand miles.

CPSIA information can be obtained
at www.ICGtesting.com
Printed in the USA
LVHW071600260122
709471LV00011B/774

9 781639 611225